MAGNETIC SENSORS - DEVELOPMENT TRENDS AND APPLICATIONS

Edited by **Aktham Asfour**

Magnetic Sensors - Development Trends and Applications
http://dx.doi.org/10.5772/66603
Edited by Aktham Asfour

Contributors

Nitu Kumar, Geetika Khurana, Anurag Gaur, Ravinder Kotnala, Ram S. Katiyar, Tianxiao Nie, Weisheng Zhao, Kang L. Wang, Dongfeng He, Alfredo García-Arribas, Eduardo Fernández, David De Cos, Yong Zhang, Rui Xuan Li, Pilar Marin, Mitra Djamal, Ramli Ramli

Published by InTech
Janeza Trdine 9, 51000 Rijeka, Croatia

Publishing Process Manager Dajana Pemac
Technical Editor SPi Global
Cover InTech Design team

First published November, 2017
Printed in Croatia
Legal deposit, Croatia: National and University Library in Zagreb

Additional hard copies can be obtained from orders@intechopen.com

Magnetic Sensors - Development Trends and Applications, Edited by Aktham Asfour
p. cm.
Print ISBN 978-953-51-3647-7
Online ISBN 978-953-51-3648-4

For EU product safety concerns:
IN TECH d.o.o., Prolaz Marije Krucifikse Kozulić 3, 51000 Rijeka, Croatia,
info@intechopen.com or visit our website at intechopen.com.

Meet the editor

Dr. Aktham Asfour is currently an associate professor at the University Grenoble Alpes (UGA), France. He received the Engineering and the PhD degrees from Grenoble INP (Institute of Engineering) in 1998 and 2002, respectively. Since 2010, he has been carrying out his research in Grenoble Electrical Engineering Lab (G2Elab). His main interests include magnetic sensors and applications as well as the systems of nuclear magnetic resonance (NMR) and magnetic resonance imaging (MRI) at very- and ultralow magnetic field, including portable and mobile systems. Between 2004 and 2010, he has been with the Grenoble Institute of Neuroscience (GIN) in Grenoble where he also carried out research in radiofrequency coils and electronics for the NMR and MRI of hyperpolarized gases.

Contents

Preface

Magnetic sensors are encountered and widely used in a large palette of applications in almost all the branches of industry and engineering. They are also key elements in a number of biomedical, space, and scientific research applications. Since a long time ago, these sensors have attracted the interest of the scientific community all around the world. This interest is still growing, and it is well justified by the will to meet new measurement challenges in all these areas and to conquer new applications.

The development of magnetic sensor is obviously closely related to the development and fabrication of new materials, which often define the final sensor performance and consequently its ability to meet the need for such application. The aim of this book is mainly to focus on some new developments and uses of materials for some types of magnetic sensors of current interest. It is organized in two sections. The first section deals with two important categories of materials including the amorphous micro-wires and the thin-film structures for sensor applications. The interest for the micro-wires is justified not only by their outstanding magnetic proprieties but also by some of their mechanical features that allow them to be used for new categories of magnetic sensors like the giant magneto-impedance (GMI) sensors, Matteucci effect–based sensors and stress sensors, etc.

A significant progress has so far been made in the fabrication of these wires. The major preparation methods, their advantages, and their disadvantages are well summarized in Chapter 1 of this book. The authors also present an overview of the general magnetic and mechanical proprieties that can be currently achieved for such wires. In Chapter 2, the author deals with an exciting and original application of amorphous micro-wires as wireless stress sensors using two important proprieties (magneto-elastic resonance and GMI). The fundamental physical principles are explained.

While these amorphous micro-wires have great potential for sensors, there are situations and applications where they could be, sometimes, not the best choice according to some considerations related, for example, to the possibility to pattern the sensitive element in micrometric shapes. This is important especially for integrated devices as it is well pointed out in Chapter 3 of the book. The authors of the chapter show that the development of thin-film structures, exhibiting GMI effect, is an emerging interesting area. Their fabrication and pattering process by photolithography is well compatible with microelectronics production. The chapter reviews the state of the art of development and characteristics of such thin-film structures, with an emphasis on the influence of the high-frequency measuring circuits on the performance.

The second section of the book is focused on materials for the magneto-resistance (MR) sensors and their applications. Simply stated, an MR sensor is based on the resistance change of the sensitive element when it is subjected to a magnetic field. This resistance change could be involved through different mechanisms such as the anisotropic magneto-resistance (AMR), the giant magneto-resistance (GMR), and tunnel magneto-resistance (TMR). This very important category of magnetic sensors is by far one of the most used in various areas like the data storage (hard disk drive) and magnetic read heads, magnetic field sensors (e.g., nondestructive testing), magnetic sensors for the measurement of nonmagnetic quantities like electrical current or angular and linear positions, as well as magnetic sensors for biomedical applications. This widespread, especially in industrial areas, is mainly stimulated by some key advantages of these MR sensors like the compatibility with the microelectronics technology, the low fabrication cost, as well as the low power dissipation.

The growing needs for these features combined with the requirement for sensitivity improvement of the MR sensors motivate the search for optimizing the existing MR technology and for the development of new materials. In this context, the authors of Chapter 4 of this book present an interesting review of the progress made in the MR effect in Mn_xGe_{1-x} structures. They show that the engineering of this material can provide transition from negative to positive MR, geometric-enhanced GMR, and electric-field–controlled MR. The well-controllable MR and the high compatibility of the process with the microelectronic technology suggest a high potential of the Mn_xGe_{1-x}-based MR sensors. In the same general context of materials development for the MR, the authors of Chapter 5 provide a detailed background and description of double perovskite Sr_2FeMoO_6 (SFMO) compound in thin films and its optimized synthesis with the will to achieve magneto-resistance proprieties at room temperature and at relatively low magnetic field.

The sixth and seventh chapters of the book deal with the implementation and applications of the GMR and AMR. Chapter 6 provides the principles of ferrite-based GMR sensors and their electronic conditioning. The authors present also a well-referenced review of several applications of these sensors.

Chapter 7 of the book addresses an important industrial application of magnetic sensors: the nondestructive testing. The author shows how to improve the performance of commercially available AMR to meet the requirements for detecting small defects in the combustion chamber of liquid rocket. The chapter also has the merit of emphasizing on the use of adequate strategies of electronic conditioning to improve the global noise performance of the sensor. In such real measurement environment, particular attention should be made to the impact of the parameters of influence, like temperature, as it is well addressed by the author.

There are obviously still many other types of magnetic sensors based on different physical principles, which are not covered by this book. The interested reader may refer to a number of works (books, research, and review papers) that detail underlying physical principles, the backgrounds, and the current state of the art of development of such sensors.

As for this book, the aim was to give a part of trends of research and development in this exciting area of science. This was only made possible thanks to the valuable work and efforts of the authors of this book who proposed high-quality chapters. I want to thank them warmly.

Aktham Asfour
University Grenoble Alpes (UGA)
Grenoble, France

Section 1

Micro-Wires and Thin-Films for Magnetic Sensors

Chapter 1

Magnetic Fibers

Rui Xuan Li and Yong Zhang

Additional information is available at the end of the chapter

http://dx.doi.org/10.5772/intechopen.70534

Abstract

The magnetic sensors based on soft magnetic effects of amorphous fibers are one of the highlights in scientific research in recent years. The amorphous fibers not only have excellent mechanical properties but also have unique magnetic properties, such as high permeability. As a result, sensors made of this kind of material can show the characteristics of sensitivity and durability. The processing and their advantages and disadvantages are mainly introduced in the chapter, and the properties reported in recent years are also summed up, including mechanical behavior, magnetic properties and shape-memory effects.

Keywords: magnetic sensor, amorphous fiber, serration behavior, glass-coated fibers, high entropy alloys

1. Introduction

In recent years, with the rapid development of information technology and automation technology, increasing requirements are put forward to the sensors, transducers and magnetic recording devices. Since the advent of amorphous fibers from 1970s, the research of these fibers has made a great progress. The amorphous fibers are defined as microfibers that have a diameter of dozens of micrometers and long-range disordered structure. This kind of fibers not only have high strength and some interesting mechanical properties [1], but they also exhibit unique magnetic effects such as large Barkhausen effect [2], Matteucci effect, giant magneto-impedance (GMI) effect [3] and giant stress-impedance effect, because of their unique internal stress distribution and magnetic domain structures [4]. As these characteristic magnetic effects are the foundation of various sensors, more and more attentions are attracted. In this chapter, four major preparation methods of amorphous fibers are focused, and the related work by the team of Y. Zhang in the University of Science and Technology Beijing (USTB) are summed up, including magnetic properties, mechanical behaviors and the shape-memory effect.

2. Preparation methods

The amorphous fibers prepared by different techniques are as follows: (1) in-rotating-water fibers, (2) glass-covered fibers and (3) melt-extracted fibers. Different preparation methods lead to different cooling rates, which in turn cause different properties. On the one hand, from outward appearance, the glass-covered fibers are different from the other two kinds of bare fibers, which can be used directly. Sometimes, the glass coatings need to be corroded for the performance tests and practical applications. On the other hand, these three kinds of fibers are different internally because of different processing techniques.

The conventional cold-drawing technique is a one-end drawing process at room temperature, where the prepared bar is drawn to pass a certain die hole. Although the surface quality can be improved through this simple process, the material selection is highly limited and the microstructure could probably be changed during this process. At present, this technology is used as a supplementary to the following three methods.

Amorphous fibers obtained by in-rotating-water quenching have been reported in 1980s. **Figure 1** shows the schematic diagram of this method. Melted by induction coil, alloys with specific composition are pressed by argon into the cooling water, which is rotating in the high-speed revolved copper mold. In this way, the fibers cool rapidly in the water, and therefore, continuous and perfect fibers are gained, which are usually smooth and have circular

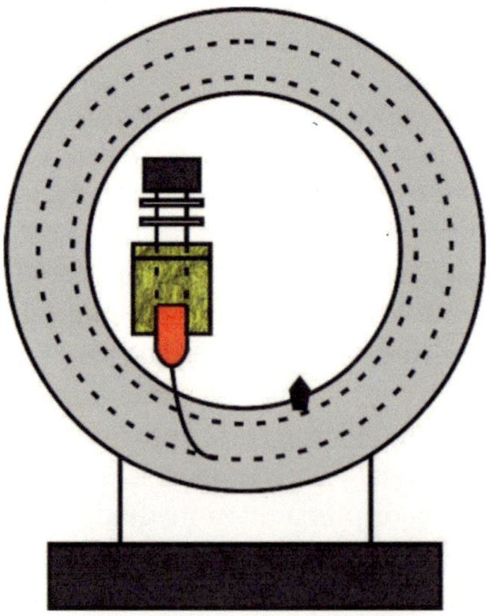

Figure 1. The schematic diagram of in-rotating-water quenching technique.

cross sections. However, the technique of producing amorphous fibers by in-rotating-water quenching is restricted greatly by the composition. This method can only be used in Fe-based, Co-based and CoFe-based alloys till now [5, 6]. The fibers also easily react with oxygen in the cooling water and thus degrade the performance. Moreover, because of the experimental conditions and the cooling rate, the diameters of these fibers are about 100 μm and thus decrease the degree of amorphization. Sometimes, these fibers should be cold drawn to decrease the diameter. Because this method is relatively simple, it develops rapidly at home and abroad and is used to produce continuous fibers at the earliest. For example, many universities developed several kinds of in-rotating-water fibers and studied their properties and technological parameters from 1990s.

A process for the production of fine metal fibers directly from the melt was first reported by Taylor in 1924 [7], and then it was improved by Ulitovsky and Wagner. As a result, this glass-coated melt spinning method is also called as Taylor-Ulitovsky method. **Figure 2** shows the schematic diagram of this method. The closed-end Pyrex glass tube and the master alloy placed in it are heated by an induction coil to the melting point. The softened glass and the liquid alloy in it are drawn out by the external force and then they are cooled down quickly by water for the purpose of quenching. Fibers with relatively small diameters are obtained through this process, and they twine around a rotating cylinder by the winding system for further use. The scanning electron microscopic (SEM) image of a glass-coated amorphous fiber is shown in **Figure 3**. The fibers are winded on a receiving coil, whose rotating speed is determinant for the fibers' diameter and the glass thickness, and hence, the fibers' properties. The diameter of the fibers ranging from 0.08 to 80 μm can be obtained. This basic method offers a route for the manufacture of metallic fibers in a single operation directly from the melt. Due to the high critical quenching rate [10^4 K/s (Kelvin temperature per second)], metastable, amorphous or supersaturated state phases can be obtained. Moreover, the coated glass

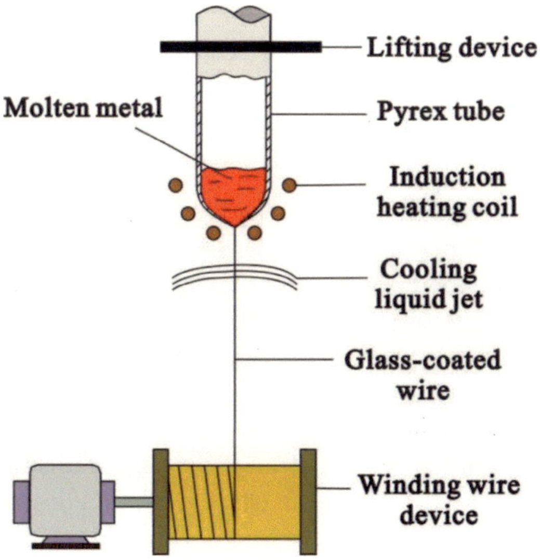

Figure 2. The schematic diagram of glass-coated melt spinning technique [12].

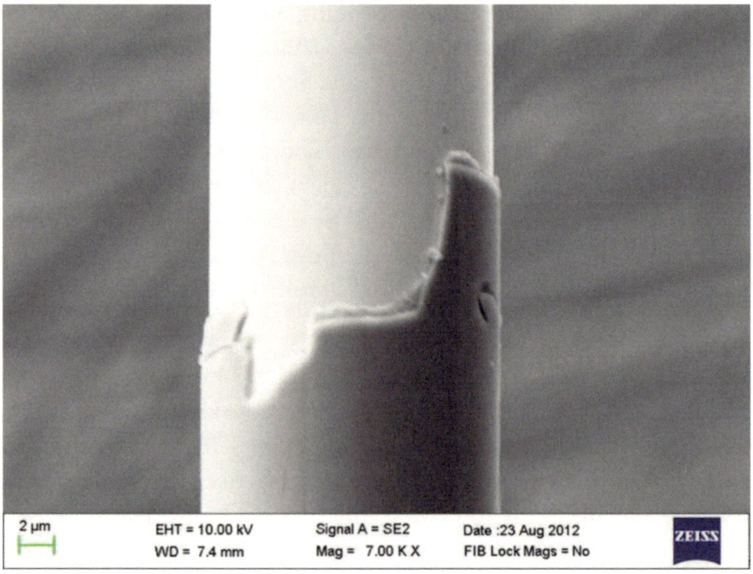

Figure 3. The SEM image of a glass-coated amorphous fiber [12].

acts as a shielding layer to improve the corrosion resistance. All these advantages contribute to promising applications such as micro electromechanical system [8]. However, the drawbacks are the limited alloy system, because this preparation method depends on the melting point of metals and the wetting between molten metal and glass. At present, this technology is relatively mature and the related research is broad in many universities [9]. Various kinds of glass-coated fibers have been obtained by the equipment (**Figure 4**) developed by our group. Our team reported Cu-15.0 at.% Sn microwires with diameters of about 184 μm [10] and $Co_{69.5}Fe_{4.35}Cr_1Si_8B_{17}$ [11] microwires with diameters of about 30 μm, which show smooth and almost flawless surface, implying that the glass-coated melt spinning method is a very effective technique in producing the high-quality microwires. In addition, Li et al. [12] prepared high-quality glass-coated $Fe_{76.5-x}Cu_1Nb_xSi_{13.5}B_9$ (x = 0, 1, 2, 3, 3.5) fibers that are also obtained by our equipment.

Put forward by Maringer at about 1974 at the earliest, the melt-extraction technique attracted much attention and was used to prepare metallic amorphous fibers at 1990s [13]. This technology is an advanced shaping methods for low dimensional materials. As shown in **Figure 5**, the working principle of this technique is as follows. Under high vacuum, the top of the alloy sample is melted by the induction heater and then extracted by the sharp edge of the rotating copper wheel. Under the combined effect of viscosity, surface tension and gravity, amorphous fibers are produced, and the cooling rate is as high as 10^6 K/s. The bare fibers with small diameters (about scores of micrometers) and a great degree of amorphization are obtained. In practice, the melt-extraction technique can be used to prepare strips, plates and fibers with different materials, the scope of which is extremely wide. This means that the unique technology overcomes the shortcomings of the three methods above. Furthermore, it has been proved that the products possess excellent electrical, magnetic and mechanical properties, as well as

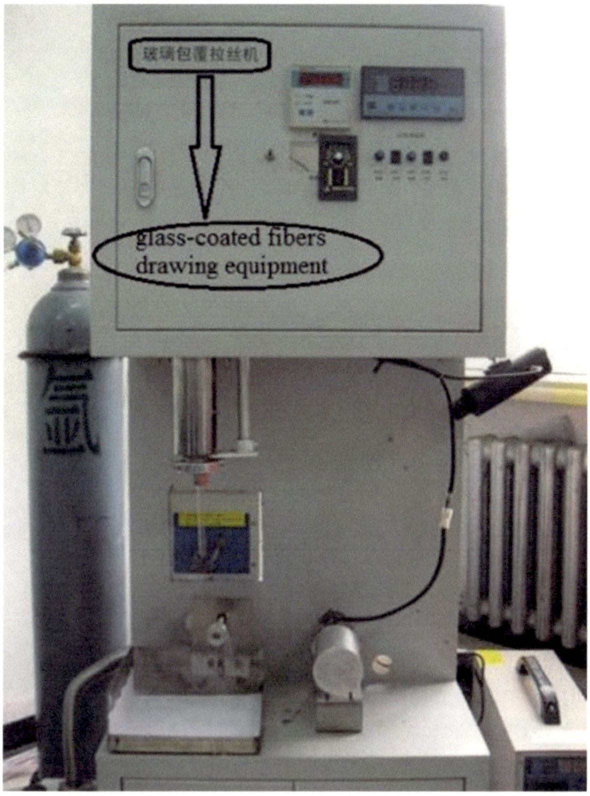

Figure 4. The picture of the glass-coated fibers drawing equipment [12].

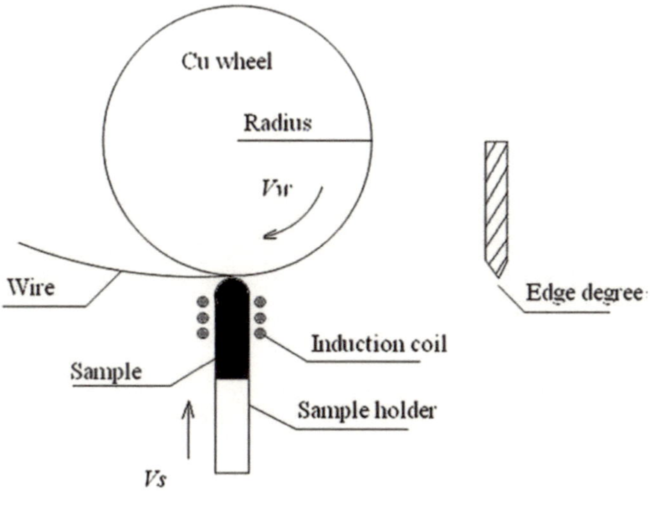

Figure 5. The schematic diagram of melt-extraction technique [14].

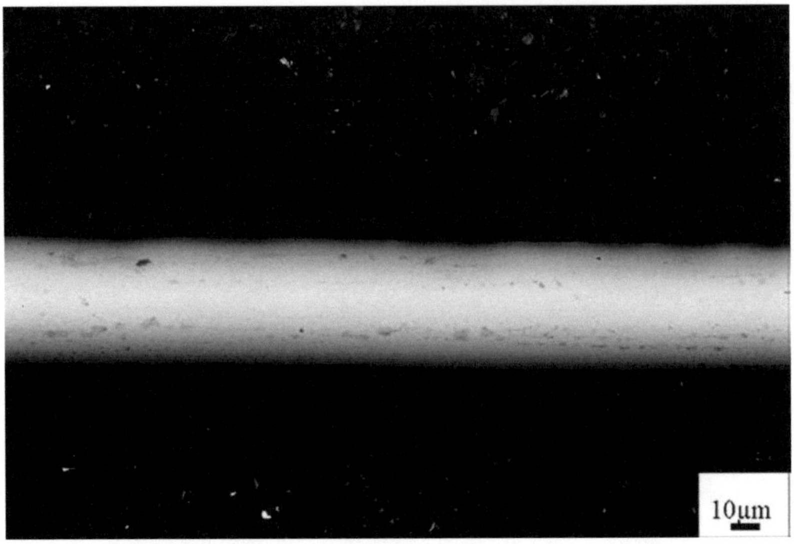

Figure 6. The SEM image of a melt-extracted fiber [14].

has high geometric symmetry. The difficulties lie in the complex preparation process. The influence factors of this process cannot be controlled exactly, and the quantitative relationship between technological parameters and the qualities is not very clear till now. At present, many universities in China are carrying out the related researches about the melt-extraction technique, and many high-performance metallic fibers have been prepared. Melt-extracted $Zr_{50.5}Cu_{27.45}Ni_{13.05}Al_9$ [14], $Co_{69.5}Fe_{4.5}Cr_1Si_8B_{17}$ [15] and Cu-Zr-Al [16] amorphous wires with almost flawless surface and precise circularity were produced by Liao (**Figure 6**), implying that the melt-extraction method was very effective in producing the high-quality wires.

3. Properties

3.1. Mechanical behavior

It is well known that the bulk metallic glasses exhibit better mechanical properties than the traditional alloys. For example, the compressive strength of CoFeTaB bulk amorphous alloys reaches up to 5185 MPa, which creates a new record of strength in metallic alloys. The amorphous fibers inherit this fine mechanical performance besides the good physical and chemical properties. In addition to the electrical and magnetic properties, the mechanical properties should be taken into special consideration to ensure a stable and reliable work. A great deal of theoretical and experimental works has indicated that the loading mode [17] and the sample size used for testing can affect the mechanical properties of amorphous alloys remarkably [18, 19]. Most of the reported metallic glass systems show considerable plasticity in compression test [20], while their tensile properties are relatively poor. Because of the normal tension stress effect [21], their tensile plasticity is poor and catastrophic fracture will happen during tensile test, which is the major limitation of the use of bulk metallic glass. Nevertheless, recent researches have shown that when the tested sample size is reduced to the nanometer scale [22, 23], the mechanical properties will become obviously different. A large plastic strain can be obtained because the shear bands no longer occur during both tension and compression. As a result, the metallic glasses with small sample sizes exhibit different deformation behaviors compared with the bulk-sized ones, which also suggests that the mechanical properties of amorphous fibers are slightly different from that of bulk amorphous alloys.

During recent years' research, the team of Y. Zhang has reported many kinds of amorphous fibers with good mechanical properties. Most of the fibers display only an elastic deformation behavior and catastrophic fracture without yielding because of their amorphous nature. Zhao et al. [24] prepared $[(Zr_{43}Cu_{50}Al_{7})_{99.5}Si_{0.5}]_{99.9}Y_{0.1}$ (Zr43) and $Zr_{50.5}Cu_{27.45}Ni_{13.05}Al_{9}$ (Zr50.5) fibers, and they all show high tensile fracture strength of 1600 MPa and a large elastic limit reaching approximately 2%. The tensile strength value increases with the increase in the content of Al. Liao et al. [14] reported $Zr_{50.5}Cu_{27.45}Ni_{13.05}Al_{9}$ alloys with fracture strength reaching 1800 MPa and elastic limit reaching 2%, and the Weibull modulus is as high as 81. The glass-coated $Fe_{76.5-x}Cu_{1}Nb_{x}Si_{13.5}B_{9}$ (x = 0, 1, 2, 3, 3.5) amorphous fibers show such a high strength as 2500 MPa, with the Weibull modulus reaching above 20, saying that the high strength is reliability [12].

It is also found by Liao that the $Co_{69.5}Fe_{4.5}Cr_{1}Si_{8}B_{17}$ [15] and Cu-Zr-Al [16] metallic glassy wires exhibit unusual nonlinear deformation behavior with irreversible elongation (**Figure 7**), and the analysis indicates that this nonlinear deformation is related to the formation of sub-nanometer voids coalesced from flow defects under tensile stress. The work suggests that the tensile failure of the glassy fibers is controlled by both the normal stress and shear stress. The dominant stress is the shear stress, and the normal stress lowers the energy barrier for the atoms to flow. The shear bands form initially on the tensile surface and then propagate toward the neutral layer.

3.2. Magnetic property

Excellent physical properties such as soft magnetic properties and unique expansibility can be obtained in amorphous materials. It has been indicated that the saturation magnetization of Fe-based amorphous alloys is more than 1.5 T, coupling with a low coercive force of about 1 A/m^2. Li reported a kind of glass-coated $Fe_{76.5-x}Cu_{1}Nb_{x}Si_{13.5}B_{9}$ (x = 0, 1, 2, 3, 3.5) fibers [12]. The results show that the magnetic behavior is greatly influenced by the structure and composition of the fibers, and it also suggests that the coercivity of the amorphous structures is far lower than that of the coarse crystal ones. Compared with all the other kinds of fibers, the glass-coated amorphous FeCuNbSiB fibers have the best comprehensive performance, which have both relatively great magnetic properties and tensile properties.

The giant magneto-impedance (GMI) effect, which is the foundation of magnetic sensors, has been found in amorphous materials in recent years. Magnetic sensors can be broadly applied in the field of geology, measuring, traffic and military, and some important parameters of the sensors include the sensitivity, stability, volume, and power consumption. The

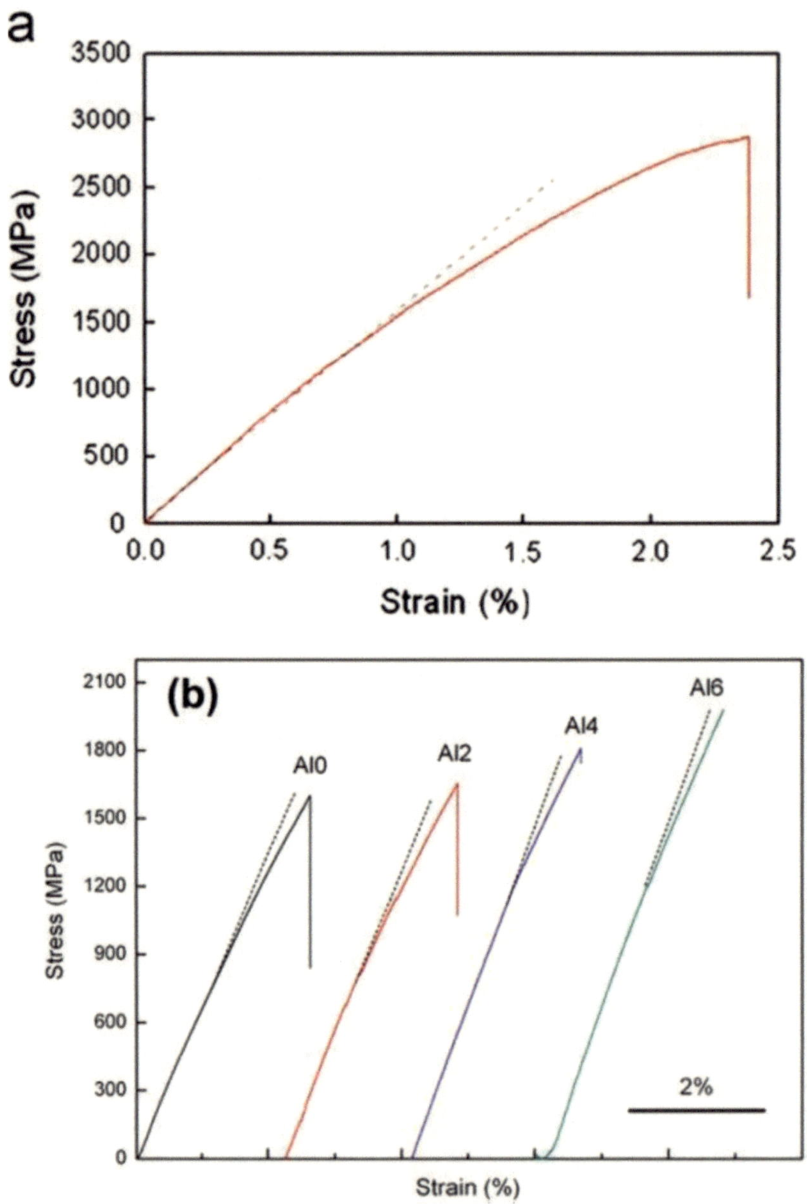

Figure 7. The unusual nonlinear deformation behavior of (a) $Co_{69.5}Fe_{4.5}Cr_1Si_8B_{17}$ [15] and (b) Cu-Zr-Al [16] metallic glassy fibers.

traditional magnetic sensors include Hall effect sensors, giant-magnetic-resistance (GMR) sensors and fluxgate sensors; while in 1994, the GMI effect was found in amorphous fibers by Professor Mohri and Panina [3], which significantly enhanced the performance of magnetic sensors. Soon afterward, the same effect was also found in other shapes, for instance, ribbon and film. But in practice, compared with ribbon samples, fibers show bigger GMI effect and sensitivity at room temperature with their highly axisymmetric structure, which meet the requirements of a number of new micromagnetic sensors.

Under the applied magnetic field, when the alternating current is flowing into the amorphous-sensitive element, there is an impedance change in it. If the voltage across the sensitive element is measured, the changes of voltage with the change of applied magnetic field can be seen. As a result, the giant magneto-impedance effect is referred to as the impedance changes along with the change of magnetic field. The GMI ratio can be expressed as:

$$\mathrm{GMI} = \frac{\Delta Z}{Z} = \frac{Z(H) - Z(H_{max})}{Z(H_{max})} \times 100\,\%, \qquad (1)$$

where Z is the impedance, ΔZ is the change of impedance, H is a certain magnetic field strength and H_{max} is the maximum value of the applied magnetic field.

The GMI effect of amorphous fibers can be attributed to the strong skin effect and the special magnetic domain structures, which are related to the stress state [4]. The internal stress usually appears during solidification and preparation, and the different thermal expansion coefficient between glass and fiber can also lead to unique internal stress states, which in turn contribute to the special domain structures. In amorphous fibers with negative magnetostrictive coefficient, the magnetic domain is axial in the outer shell and circumferential in the inner core, while the domain, which is in the inner core, will change into radial after the glass coat is applied, because of the increased axial stress, as is shown in **Figure 8**.

As indicated above, due to the circular magnetic anisotropy caused by the inner stress and magnetostrictive effect, the annular magnetic domain is formed on the surface. The passed electric current creates an easy axis, which lead to the movement of domain wall and the annular magnetization, while the applied longitudinal magnetic field prevents the change of the annular magnetic flux. As a result, the transverse magnetic permeability varies sensitively with changes in the external magnetic field. Then, the varying transverse magnetic permeability acts

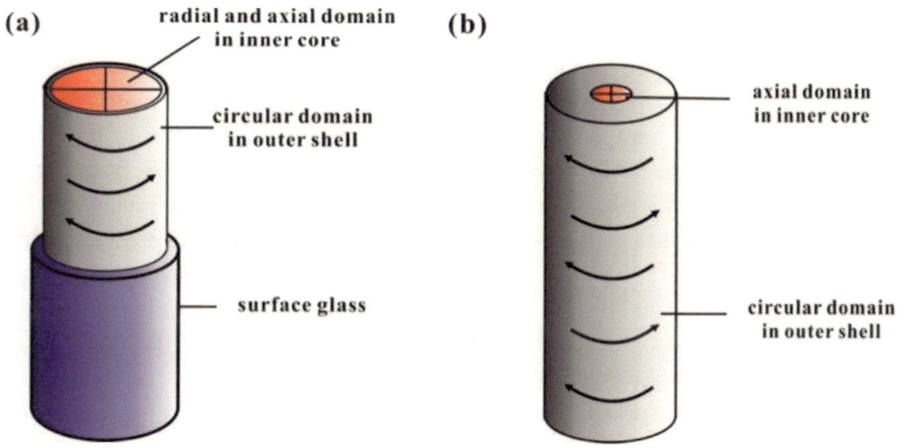

Figure 8. Magnetic domain distribution of (a) as-cast and (b) glass-removed fibers [12].

as a promotion to the changes in the impedance. This is because the impedance is a function of skin depth, whose mathematical expression is

$$\delta_m = \sqrt{2\varrho/\mu\omega}, \tag{2}$$

where ϱ is the electrical resistivity, μ is the transverse magnetic permeability and ω is the angular frequency of the alternating current. Finally, the changed rules of external magnetic field are obtained from the changes of impedance.

In the aspect of material, the magnetostrictive coefficient of Co-base amorphous fibers is -10^{-7}, approximately thought of as zero, and this kind of fibers has high corrosion resistance, tensile strength and domain wall energy density, plus special magnetic domain structures, and thus, Co-base amorphous material is a kind of representative GMI materials [25]. The sensitivity of the giant magneto-impedance effect is affected by various factors, such as material types, the number and diameter of fibers and the glass coat. García et al. [26] discovered that in $Co_{59.2}Fe_{4.8}Ni_{11}B_{14}Si_{11}$ amorphous fibers, the GMI ratio increases first and then decreases with the increase in fibers under high-frequency current of about 500 MHz, whereas the value of ratio decreases gradually under low-frequency current. Zhang et al. [27] studied the influence of diameter on GMI ratio. The results show that the GMI ratio first increases and then decreases with the increase in diameter, and the coated glass also has effect on the GMI ratio. Zhao et al. [11] suggests that the cold drawing has a great influence on the GMI effect. A transition from single-peak (SP) to double-peak (DP) GMI behavior is observed in $Co_{69.5}Fe_{4.35}Cr_1Si_8B_{17}$ microwires after cold drawing, which is caused by the induced circumferential stress. This circumferential stress appears during cold drawing, and it results in the significant changes of the surface magnetic anisotropy, which changes from helical domains to circular domains. The GMI ratio as well as field sensitivity is improved significantly after cold drawing and that would be significant for application.

3.3. Shape-memory effect

In addition, a significant shape-memory effect and superelasticity were found in Y. Zhang's research group. They prepared the Cu-15.0 at.% Sn microwires by glass-coated melt spinning and reported that the bent sample, which were deformed to "U" shape in liquid nitrogen, remembered its initial shape and recovered by itself at room temperature. Standing in sharp contrast to the slightly shape-memory effect in single crystal of Cu-15.0 at.% Sn alloys, this phenomenon can be explained by the high cooling rates suppressing the precipitation of other phases, and thus very good shape-memory effects are obtained. Furthermore, the plasticity of these kinds of fibers is great and the fracture strain reaches 14.2%, while it is usually brittle when the fibers are in the polycrystalline state. From a long-term development, the experiments and results will contribute to solving the limitation of Cu-based shape-memory alloys, which will also enlarge the application range of the Cu-based alloys and promote the development of shape-memory alloys.

4. Conclusion

The main preparation methods, as well as their advantages and disadvantages, are mainly introduced in this paper, and the properties reported in recent years are also summed up, including good mechanical properties, magnetic properties and shape-memory effects. It now appears that the preparation technology of amorphous fibers is relatively mature, and we can choose the most appropriate method according to the basic characteristics of the alloy system and our requirement. In terms of the properties and applications, the soft magnetic properties and special magnetic effects of the fibers have been fully studied and successfully applied to the magnetic sensors, achieving the goals of miniaturization. However, other performance such as shape-memory effect has not been applied in practice, so lots of work in this aspect need to be carried out.

Acknowledgements

Y. Zhang would like to thank the financial support from National Natural Science Foundation of China (NSFC), with grant nos. 51471025 and 51671020.

Author details

Rui Xuan Li and Yong Zhang*

*Address all correspondence to: drzhangy@ustb.edu.cn

State Key Laboratory for Advanced Metals and Materials, University of Science and Technology Beijing (USTB), Beijing, China

References

[1] Wu Y, Wu HH, Hui XD, Chen GL, Lu ZP. Effects of drawing on the tensile fracture strength and its reliability of small-sized metallic glasses. Acta Materialia. 2010;**58**(7):2564-2576

[2] Gonzalez J, Murillo N, Larin V, Barandiaran JM, Vazquez M, Hernando A. Magnetic bistability of glass-covered Fe-rich amorphous microwire: Influence of heating treatments and applied tensile stress. Sensors and Actuators A: Physical. 1997;**59**(1-3):97-100

[3] Panina LV, Mohri K. Magneto-impedance effect in amorphous wires. Applied Physics Letters. 1994;**65**(9):1189-1191

[4] Chiriac H, Ovari TA. Amorphous glass-covered magnetic wires: Preparation, properties, applications. Progress in materials Science. 1996;**40**(5):333-407

[5] Hagiwara M, Inoue A, Masumoto T. Mechanical properties of Fe-Si-B amorphous wires produced by in-rotating-water spinning method. Metallurgical Transactions A. 1982;**13**(3):373-382

[6] Inoue A, Chen HS, Krause JT, Masumoto T, Hagiwara M. Young's modulus of Fe-, Co-, Pd- and Pt-based amorphous wires produced by the in-rotating-water spinning method. Journal of Materials Science. 1983;**18**(9):2743-2751

[7] Taylor GF. A method of drawing metallic filaments and a discussion of their properties and uses. Physical Review. 1924;**23**(5):655

[8] Chiriac H, Colesniuc CN, Ovari TA, Ticusan M. In situ investigation of the magnetization processes in amorphous glass-covered wires by ferromagnetic resonance measurements. Journal of Applied Physics. 1999;**85**(8):5453-5455

[9] Wang RL, Ruan JZ, Kong XH, Yang W, Wang JT, Zhao ZJ. Giant magneto-impedance on glass-coated microwires with coper layer. Journal of East China Normal University (Natural Science). 2013;**1**:115-120

[10] Zhao YY, Li H, Wang YS, Zhang Y, Liaw PK. Shape memory and superelasticity in amorphous/nanocrystalline Cu-15.0 atomic percent (at.%) Sn wires. Advanced Engineering Materials. 2014;**16**(1):40-44

[11] Zhao YY, Hao HH, Zhang Y. Preparation and giant magneto-impedance behavior of Co-based amorphous wires. Intermetallics. 2013;**42**:62-67

[12] Li RX, Hao HH, Zhao YY, Zhang Y. Weibull statistical reliability analysis of mechanical and magnetic properties of FeCuNbxSiB amorphous fibers. Metals. 2017;**7**(3):76

[13] Inoue A, Amiya K, Yoshii I, Kimura HM, Masumoto T. Production of Al-based amorphous alloy wires with high tensile strength by a melt extraction method. Materials Transactions, JIM. 1994;**35**(7):485-488

[14] Liao WB, Hu JM, Zhang Y. Micro forming and deformation behaviors of $Zr_{50.5}Cu_{27.45}Ni_{13.05}Al_9$ amorphous wires. Intermetallics. 2012;**20**(1):82-86

[15] Liao WB, Chen ZD, Li M, He JP, Zhang Y. Nonlinear tensile deformation behavior of melt-extracted $Co_{69.5}Fe_{4.5}Cr_1Si_8B_{17}$ amorphous wires. Materials Letters. 2013;**97**:195-197

[16] Liao WB, Zhao YY, He JP, Zhang Y. Tensile deformation behaviors and damping properties of small-sized Cu-Zr-Al metallic glasses. Journal of Alloys and Compounds. 2013;**555**:357-361

[17] Schuh CA, Lund AC. Atomistic basis for the plastic yield criterion of metallic glass. Nature Materials. 2003;**2**(7):449-452

[18] Wu FF, Zhang ZF, Mao SX. Size-dependent shear fracture and global tensile plasticity of metallic glasses. Acta Materialia. 2009;**57**(1):257-266

[19] Jang DC, Greer JR. Transition from a strong-yet-brittle to a stronger-and-ductile state by size reduction of metallic glasses. Nature Materials. 2010;**9**(3):215-219

[20] Liu YH, Wang G, Wang RJ, Zhao DQ, Pan MX, Wang WH. Super plastic bulk metallic glasses at room temperature. Science. 2007;**315**(5817):1385-1388

[21] Zhang ZF, Eckert J, Schultz L. Difference in compressive and tensile fracture mechanisms of $Zr_{59}Cu_{20}Al_{10}Ni_{8}Ti_{3}$ bulk metallic glass. Acta Materialia. 2003;**51**(4):1167-1179

[22] Guo H, Yan PF, Wang YB, Tan J, Zhang ZF, Sui ML. Tensile ductility and necking of metallic glass. Nature Materials. 2007;**6**:735-739

[23] Volkert CA, Donohue A, Spaepen F. Effect of sample size on deformation in amorphous metals. Journal of Applied Physics. 2008;**103**(8):083539

[24] Zhao YY, Hu JM, Zhang Y. Processing and properties of CuZr-based amorphous microwires. Procedia Engineering. 2012;**36**:551-555

[25] Mohri K, Kawashima K, Kohzawa T, Yoshida H. Magneto-inductive element. IEEE Transactions on Magnetics. 1993;**29**(2):1245-1248

[26] García C, Zhukova V, Zhukov A, Usov N, Ipatov M, Gonzalez J, Blanco JM. Effect of interaction on giant magnetoimpedance effect in a system of few thin wires. Sensor Letters. 2007;**5**:10-12

[27] Zhang ZH, Li BY, Cui WL, Xie JX. Influence of glass coating thickness and metallic core diameter on GMI effect of glass-coated $Co_{68}Fe_{4.5}Si_{13.5}B_{14}$ amorphous microwires. Journal of Magnetism and Magnetic Materials. 2011;**323**(12):1712-1716

Chapter 2

Wireless Stress Sensor Based on Magnetic Microwires

Pilar Marín

Additional information is available at the end of the chapter

http://dx.doi.org/10.5772/intechopen.70354

Abstract

The development of wireless sensors and biosensors is a topic of great current interest. Amorphous magnetoelastic microwires are perfect candidates to be used as sensing elements based on two important properties, that is, magnetoelastic resonance and high-frequency giant magnetoimpedance. It was observed that such microwires present the key feature of performing magnetoelastic resonance, at the kHz range of frequency, in the absence of applied field. This fact, in addition to their small size, gives the microwires unique advantages over the widespread ribbons, currently in use as magnetoelastic sensors. The frequency, amplitude, and damping of the vibration give information of the sensor environment. On the other side, the microwire reflectivity in the microwave range can be modulated by means of magnetoimpedance effect. The maximum-induced electric current, as well as the maximum ac modulation, occurs for frequencies determined by the microwire length. The modulation also varies as a function of the dc-applied field and applied stress.

Keywords: magnetostriction, magnetoelastic resonance, high-frequency giant magnetoimpedance, amorphous magnetic microwire, biomedical sensors

1. Introduction

In recent years, much interest and effort have been devoted to develop soft magnetic materials due to their technological potential [1]. The main use of these materials can be found in the sensing industry which includes a broad spectrum of applications ranging from the automotive, mobile communication, chemistry, and biochemistry industry among many others [2–5]. Amorphous microwires are one of the most widely studied soft materials. They are fabricated by means of extracting melt-spinning Taylor technique [6]. Those microwires are composed by a metallic core and a Pyrex cover both in the micrometer range. The metallic core provides the magnetic behavior, while the cover has a protective and stress-inducting function [7]. The ratio between the total diameter and the magnetic core, often called aspect ratio, is one of the key parameters of such microwires, since magnetic properties depend dramatically on it. For high values of the aspect ratio, the microwires present a bistable hysteresis loop, while this bistability vanishes for low ones [8, 9]. In fact, amorphous or nanocrystalline magnetic microwires are among the softest materials. Many properties of these materials have been deeply studied both from the point of view of the basic physics and the applications [10, 11]. This is the case of the giant magnetoimpedance effect [12], bistability [13], ferromagnetic resonance [14], and magnetoelastic resonance [15, 16]. It is easy, also, to find much literature regarding microwave-related applications of microwires or microwire-based materials [17–19]. Some of these articles have nicely shown how different arrangements of microwires forming arrays or embedded in different types of matrices may be used for enhancing their sensitivity as GMI elements or electromagnetic waves absorbing materials [20].

In the frequency range of GHz, some experimental and theoretical studies of the effect of the magnetization on the scattering properties of a single microwire have been recently developed [21]. This kind of work gives experimental evidence showing that the microwave scattering by a single microwire depends on the magnetic permeability with sufficient strength to be experimentally detected as an effect of the GMI. Furthermore, this dependence was also used to show the potential of such microwire as a wireless field and/or stress sensor. These experimental results are followed by a theoretical approach where the influence of the microwire magnetic state in its microwave reflection features is taken into account. Based on these studies, further experimental work shows an application of such microwire as a wireless stress sensor with the particular application of pressure detection in a hydraulic circuit simulating cardiovascular conditions [22].

Besides these investigations on magnetic microwires, it should be stated that technological development has spurred the growing interest in the investigation of new biosensors aimed at simplifying present-day diagnostic methods and thereby improving medical care, so that it improves the quality of life of the patients and allows for outpatient treatment for a number of pathologies, avoiding unnecessary hospital admissions [23]. Magnetic sensors are at the helm of technological development seen in this field over the last decades, offering numerous advantages attributed to their elevated sensitivity, reduced size, systems without the need for an external source of energy, and wireless connections. The use of wireless sensor network (WSN) technologies offers the possibility of developing implantable biomedical sensors allowing for the monitorization and follow-up of certain physiological parameters with precise and, up until now, unthinkable measurements. Based on magnetoelastic character or microwires, it has been possible to develop a wireless magnetic sensor for postoperative follow-up procedures of vascular surgery [24].

The aim of the present paper is to show the physical fundamentals and the particular applications of magnetic microwires as stress sensors

2. Scattering of electromagnetic waves by a single microwire

2.1. Magnetostrictive magnetic microwires

A magnetic microwire is a filament with an amorphous structure, whose nucleus is composed of an alloy of metals, the most frequent being iron and cobalt, with a Pyrex covering as insulant, manufactured by means of an ultrafast cooling process resulting in microwires with a maximum total diameter around 100 m. Coils containing around 2 or 3 km of wire with a weight around 1–2 g are tailored by Taylor-Ulitovsky technique [25]. **Figure 1a** shows a photograph corresponding to a microwire scanning electron microscope (SEM) image. The composition is $Fe_{2.25}Co_{72.75}Si_{10}B_{15}$, and corresponding dimensions are a total diameter of 100 m and a metal nuclei diameter around 80 m.

Figure 2 shows a schematic diagram of the method used by Taylor-Ulitovsky. It consists of the rapid drawing of a softened glass capillary in which the molten metal is entrapped. The capillary is drawn from the end of a glass tube containing the molten alloy. Previously, a metallic pellet of the master alloy, prepared by induction melting of pure elements, has been placed inside the sealed end glass tube; then, the alloy is melted by a high-frequency field of an inductive coil, and the end of the glass tube is softened. Hence, around the molten metal drop, there is a softened glass cover which allows the drawing of the capillary. A low level of vacuum within the glass tube prevents metal oxidation. It also assures stable melt-drawing conditions, in conjunction with induction heating, employing the levitation principle. In order to ensure continuity of the process, there is a glass tube displacement with a uniform feed-in speed. The rapid cooling rate required to obtain an amorphous structure is between 10^5 and 10^6 Ks^{-1}.

Amorphous magnetoelastic Co-based magnetic microwires with negative but low magnetostriction have been used for conducting this study related with wireless stress sensing technologies. To ensure soft magnetic properties, cobalt-rich compositions, that is, $Fe_{2.25}Co_{72.75}Si_{10}B_{15}$, should be adjusted to obtain a nearly zero but negative magnetostriction constant. The

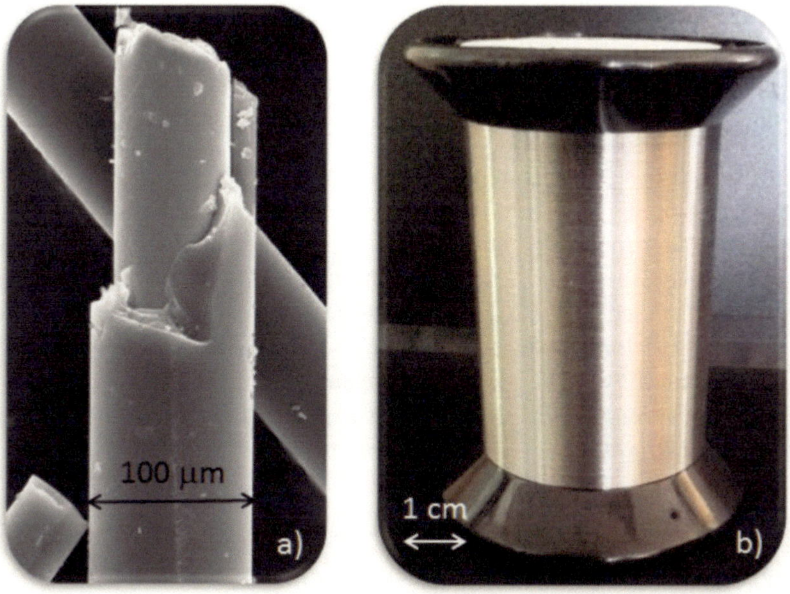

Figure 1. SEM image of an amorphous is $Fe_{2.25}Co_{72.75}Si_{10}B_{15}$ microwire (a) and photograph of coil of casted wire containing 2 g of material (b).

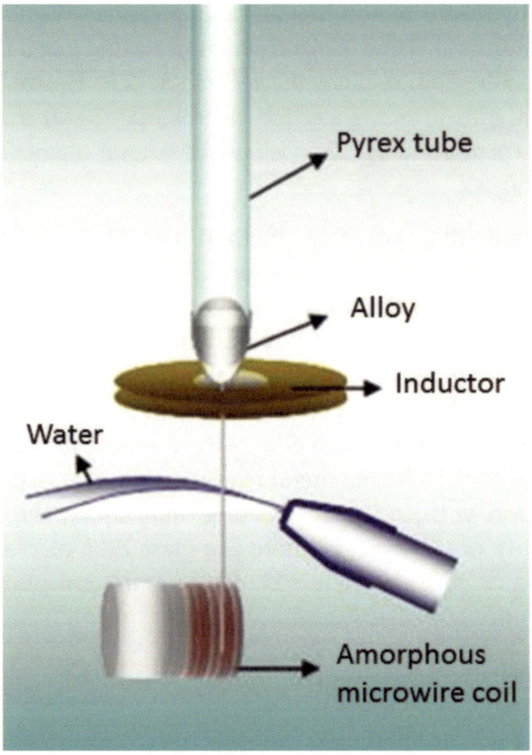

Figure 2. Microwire casting method.

bamboo-like magnetic domains observed in this type of wires have a circumferential magnetization that yields in a typical soft magnetic material hysteresis loop that exhibits low coercivity and large permeability levels [26].

Two unique characteristics which convert negative low magnetostrictive magnetic microwires into excellent sensor elements should be highlighted. On the one hand, their high magnetostriction, together with their low anisotropy, makes them extremely sensitive to small changes in mechanical stress and that such changes be translated into changes in their magnetization when subjected to an external magnetic field. In addition, because of its high magnetic susceptibility along with its diameter in microns, it is able to modulate a high-frequency signal emitted by an antenna. Thus, changes in stress or pressure of a fluid will cause a variation of the mechanical stress on the sensor, which will involve a variation of its magnetization and the emitted wave that will be detectable wirelessly through a receiving antenna [27].

Microwires' magnetic hysteresis loops are easily detected using experimental setups consisting of a primary and a secondary coil by a conventional low-frequency induction technique. The stress influence on microwire magnetic behavior can be quantified by means of hysteresis loop evolution when tensile stresses are applied to the microwire by means of hanging different weights. **Figure 3** shows the evolution of $Fe_{2.25}Co_{72.75}Si_{10}B_{15}$ hysteresis loops with tensile

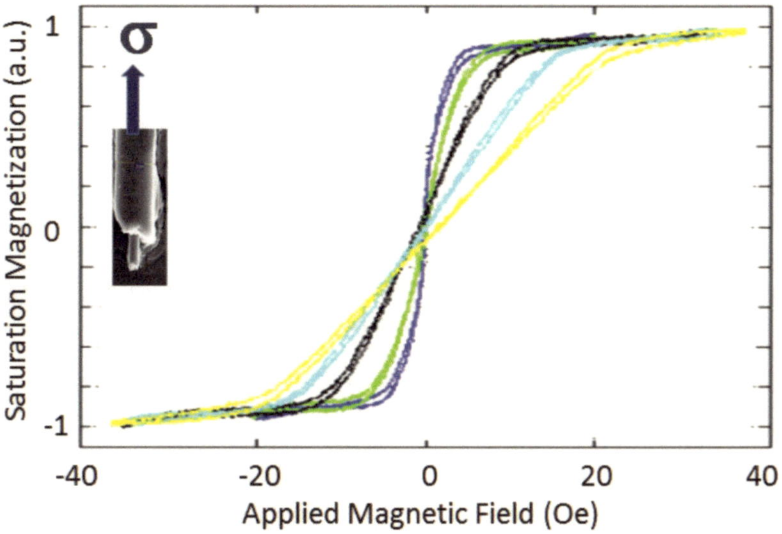

Figure 3. Stress influence on microwire hysteresis loop: 0 (■), 112 (■), 345 (■), 580 (■), and 813 (■) MPa.

stress, from 0 to 813 MPa. Magnetostrictive character of magnetic microwires is observed by means of a clear decrease of magnetic susceptibility with increasing tensile stress.

2.2. Tuned scattering of electromagnetic waves by a finite length ferromagnetic microwire

The metallic nature of ferromagnetic microwires determines the characteristics of their interaction with electromagnetic waves. An alternating electric field E_{inc}, with frequency ω, that is, at the Gigahertz range, and parallel to the axis of the wire and uniform along its length, excites the microwire inducing an electric current I(z) along the microwire. The total electric field E, at any point of the space, can be obtained as the sum of both contributions, that is, $E = E_{inc} + E_{sca}$, where E_{sca} is the electric field generated by the induced current in the thin wire. Following the antenna theory, the electric current can be computed by the Hallen equation [28]:

$$2i\int_{-\frac{L}{2}}^{\frac{L}{2}}\left(Z_0\frac{e^{-ik|r-r'|}}{4\pi|r-r'|}+Z_ie^{-ik|z-z'|}\right)Idz' = C_1sinkz + C_2coskz + 2\frac{E_0}{k} \tag{1}$$

where Z_0 the medium impedance, C_1 and C_2 are arbitrary constants fixed by the boundary conditions I(L/2) = I(−L/2) = 0, k is the wave vector, and E_0 is the amplitude of the microwave electric field. The wire impedance per unit length, Z_i is given by Eq.(2). It depends on the magnetic permeability of the microwire according to its general expression [29]:

$$Z_i = \sqrt{\frac{\omega\mu}{2\sigma}}\frac{(1-i)J_o\left(\frac{(1-i)a}{\delta}\right)}{2\pi a J_1\left(\frac{(1-i)a}{\delta}\right)} \tag{2}$$

where J_o and J_1 are the first-kind Bessel functions, $\delta = \sqrt{2/\omega\mu\sigma}$ is the magnetic skin depth [30], $2a$ is a characteristic cross-sectional size, and σ and μ hold for the electrical conductivity and the magnetic permeability of the microwire, respectively. It is important to remark that although the magnetic permeability of a microwire is strictly a tensor magnitude [30–32], for simplification and as an approximation, the magnetic permeability was taken as a scalar magnitude [33]. A more general treatment about the relation among the magnetic permeability and the electric impedance of a microwire can be seen in Refs. [34, 35]. Nevertheless, the most important issue related to the induced current in a magnetic microwire holds in the Z_i expression, given by Eq. (2) and its own dependence on the magnetic permeability that can be dramatically changed in ferromagnetic microwires by applying a bias magnetic field [36].

In order to quantify the total induced current, I(z), in magnetic microwires in the presence of microwaves verifying the integral equation, Hernando et al. [37] have applied a pulse function point matching method [35] for microwires with radius a = 33 μm, electric conductivity of order of 10^7 S, relative magnetic permeability value of 2, and an amplitude of the electric field about 1 V/m. As previously mentioned, the boundary conditions were applied on the electric current to set the values of the constants C_1 and C_2. Furthermore, the effect produced in the induced current by variations of the microwire magnetic permeability at different microwave frequencies was also studied. **Figure 4** shows the modulus $I(z)$ at the center of different microwires, $I(0)$, with lengths of 5, 10, 15, and 30 cm as a function of the microwave frequency for different values of the magnetic permeability.

The results clearly show a current distribution typical for thin metallic wires, where the electric current is maxima at certain frequencies depending on the length of the wire: dipole antenna resonances. The maximum $I(0)$ variation induced by changes of the magnetic permeability is shown to correspond to that obtained at the dipole antenna resonances. Consequently, maximum giant magnetoimpedance effect will be achieved at these resonance frequencies.

In order to corroborate the proposed theoretical model, microwave characterization of 33 μm radius Co-based microwires and different lengths was carried out in the 500 MHz–2 GHz frequency range. The microwave scattering properties of these magnetic microwires were experimentally analyzed, as shown in **Figure 5**, by using two helical antennas (with 21.5 cm of length, 3 cm of radius, central frequency of 1 GHz, and a wave band of 500 MHz) connected to a Programmable Network Analyzer from Agilent Ltd., working in the frequency range comprised between 10 MHz and 20 GHz. Furthermore, a microwave polarizer was used to ensure that the electric field of the microwave was parallel to the axis of the microwire. The experiments were based on S_{ij} parameter measurements defined as shown in Eq. (3) where P_1 and P_2 stand for the input and output signal, respectively:

$$S_{ij} = 20log_{10}\frac{P_i}{P_j}; i, j = 1, 2 \tag{3}$$

In our case, we restricted the study to the S_{21} parameter. The choice was based in the interesting physical meaning, since it represents the measured power in the receiving antenna over the outgoing power in the emitting one.

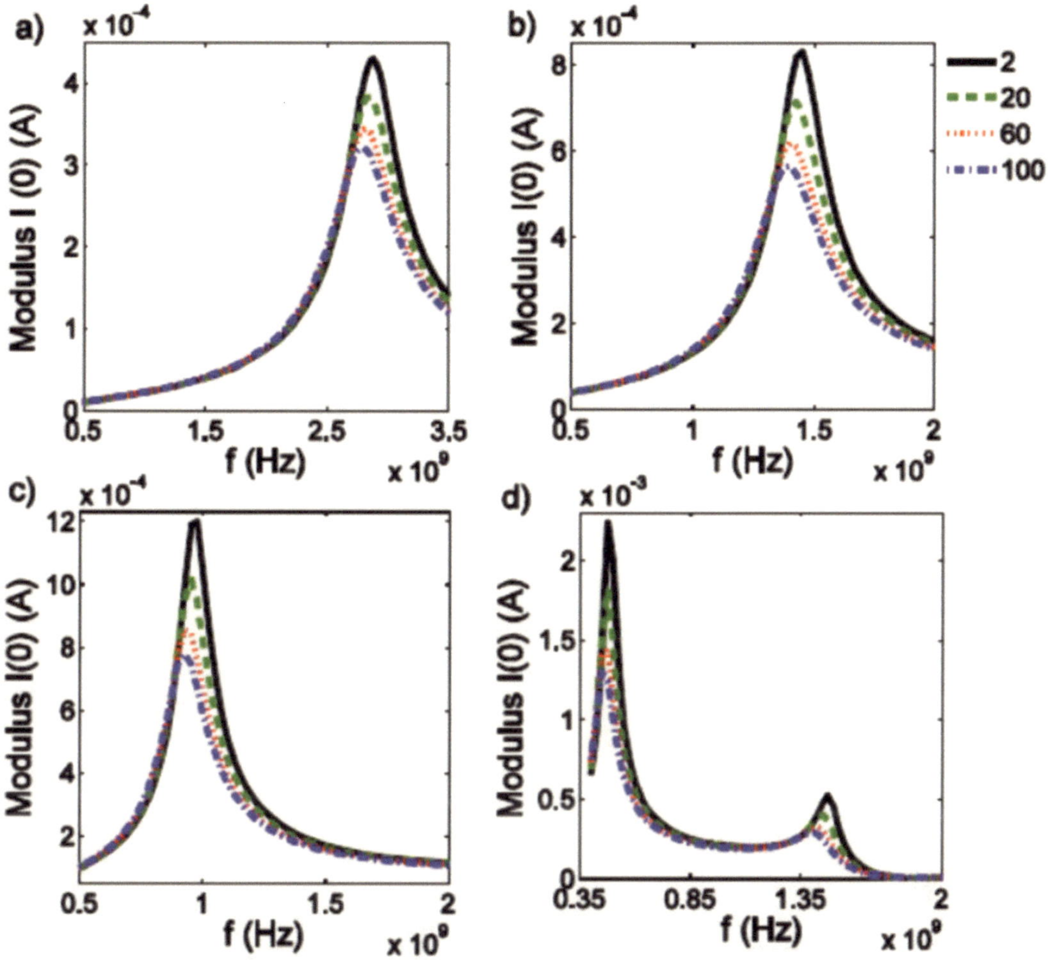

Figure 4. Frequency spectrum of the modulus of the electrical current at the center of a microwire for different relative, permeabilities for (a) 5 cm, (b) 10 cm, (c) 15 cm and (d) 30 cm microwire length where 2, 20, 60, and 100 are values of permeability. (Figure published with copyright permission from IEEE Transactions on Antennas and Propagations. 2016;64(3):1112-1115).

It is important to take into account that the scattering coefficient is directly related to the electrical current induced along the microwire by the incident wave since the scattered wave is generated by the electric current along the microwire (1) (the field distribution corresponds for a dipole antenna [35]).

As previously demonstrated (**Figure 4**), the electric current can be modified changing the magnetic permeability of the ferromagnetic microwire. For this reason, two Helmholtz coils were used to apply an ac-bias magnetic field, parallel to the axis of the microwire, with amplitude of 2.7 Oe and a frequency of 10 Hz. The experiments were performed by applying a dc magnetic field in addition to the abovementioned ac-bias magnetic field, also parallel to the axis of the microwire. Following this procedure, the modulation is driven around different points of the hysteresis loop or different magnetic states. The ac field also gives rise to a

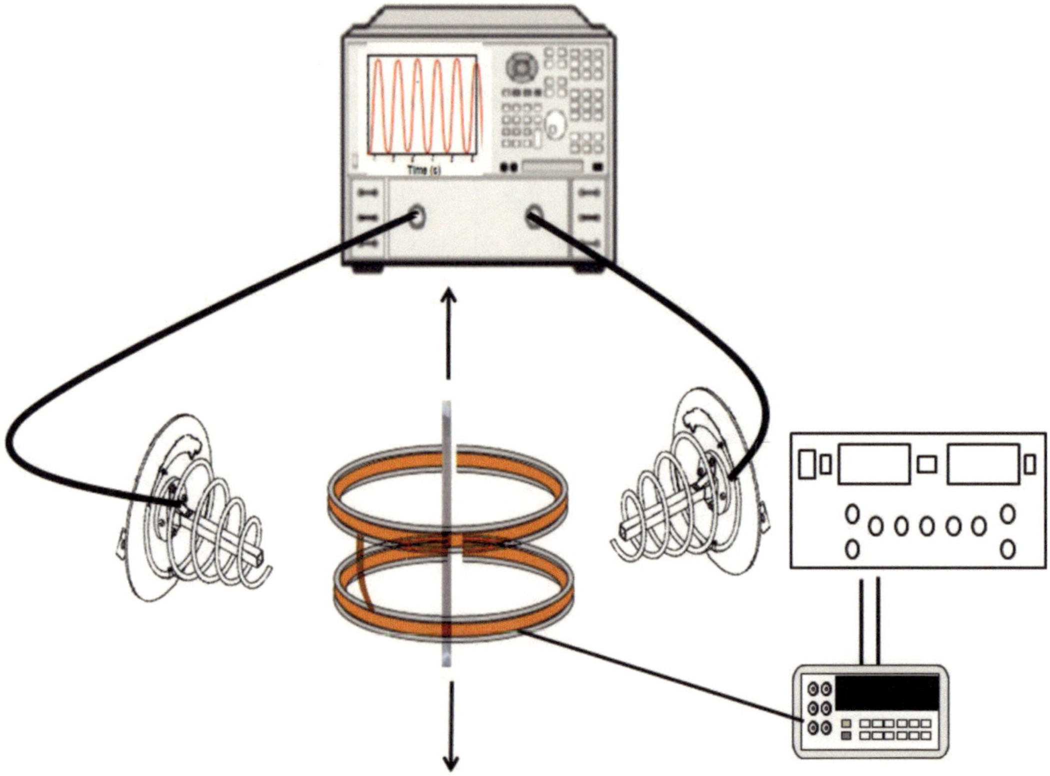

Figure 5. Experimental setup for combined application of high-frequency electromagnetic wave with frequency between 10 MHz and 20 GHz by means of helicoidally antennas connected to Programmable Network Analyzer and 10 Hz low-frequency bias field using Helmholtz coils.

modulation of the microwire permeability that is reflected as a modulation in the scattering coefficient: ΔS21. The amplitude of the modulation of the scattering coefficient due to the ac-bias field as a function of the dc-applied field for a 10 and 15 cm microwire length is represented in **Figure 6**.

In the insets of **Figure 6**, the microwave spectrum for each microwire length at zero dc field is depicted. For each case, a maximum variation of the amplitude of the modulation induced in the scattering coefficient at the following frequencies, 1.4 GHz for the 10 cm length (**Figure 6a**) and 1 GHz for the 15 cm microwire length (**Figure 6b**), is observed. This experimental feature perfectly matches with the antenna resonance frequencies deduced by numerical calculations as shown in **Figure 4**.

At constant frequency the scattering coefficient modulation increases as the dc-applied field, reaching a maximum, at 3 Oe for both microwires length. This field corresponds to the experimental coercive field measured at 10 Hz by typical induction measurements (see **Figure 7**) [27].

Effectively, the maximum permeability variation induced by an applied field is expected to occur when the latter equals the coercivity of the wire. For dc magnetic fields larger than *3 Oe*,

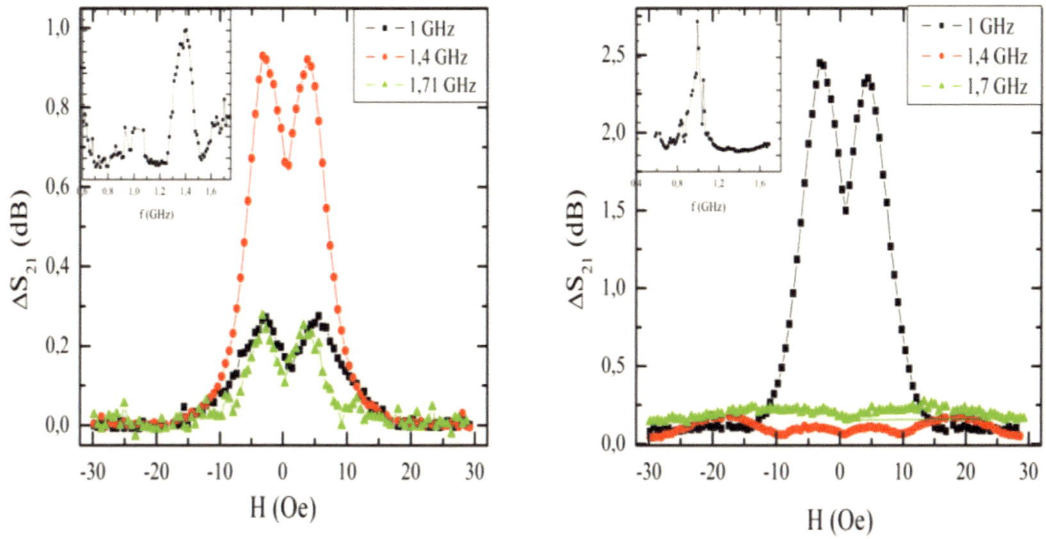

Figure 6. Relative variation of the scattering coefficient S_{21} as a function of the applied dc-bias field for (a) a 10-cm microwire length where the maximum variation occurs at 1.5 GHz and (b) a 15-cm-microwire length where the maximum variation occurs at 1 GHz. In the insets the microwave spectrum for each microwire at zero dc-bias field is represented [37] (Figure published with copyright permission from IEEE Transactions on Antennas and Propagations. 2016;64(3):1112-1115).

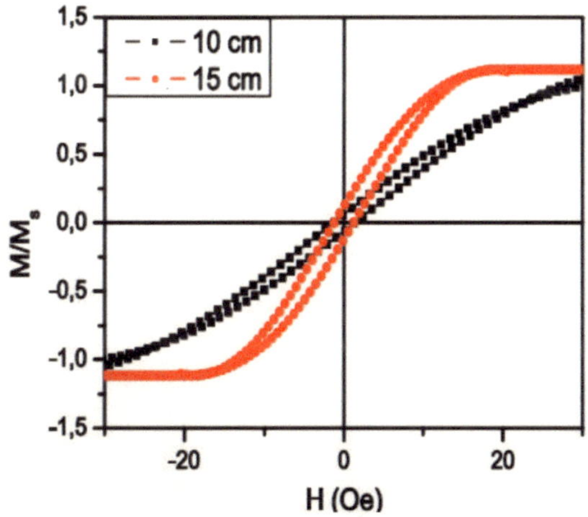

Figure 7. Hysteresis loops for Co-rich microwires with negative magnetostriction at 10 Hz measured by ac induction technique with 10 and 15 cm length, respectively (adapted from Hernando et al. [37]).

the modulation of the scattering field diminishes and finally vanishes for dc fields well above the saturation field (the impedance, Z_i, of the magnetic microwire is not modulated because of the magnetic permeability that remains constant). Therefore, it has been shown in the experiments that the maximum changes in the microwave spectrum by the GMI occur at the antenna resonances as was shown applying the antenna theory. The curves shown in **Figure 6** suggest that at the dipole resonance the system can be used as a sensitive detector of magnetic fields. This effect can be used for different sensing applications.

2.3. Stress and field contactless sensor based on the scattering of electromagnetic waves by a single ferromagnetic microwire

The following paragraphs [38] present experimental evidence that the microwave scattering by a single microwire depends on the magnetic permeability with sufficient strength to be experimentally detected as an effect of the GMI. Furthermore, this dependence has been also used to show the potential of such microwire as a wireless field and/or **stress sensor**.

A 10-cm-length sample of magnetostrictive $Fe_{2.25}Co_{72.75}Si_{10}B_{15}$ amorphous microwire, presenting high magnetoimpedance effect, of metallic core radius of a = 33 μm and total radius, including the Pyrex outer shell, of 50 μm was placed between two helical antennas working in the GHz range with circular polarized radiation as shown in **Figure 5**. As previously stated in the present work, due to its amorphous structure, even though they exhibit low or moderate magnetostriction, highly sensitive magnetoelastic elements for detecting applied stresses are thoroughly discussed elsewhere since long time ago [39].

Figure 8 displays the experimental result of measuring the S_{21} parameter over time for the aforementioned setup with an emission frequency of the antennas set at $f_{antennas}$ = 1.2 GHz. The modulation of the S_{21} parameter shown in **Figure 8** is produced by the application of a bias field with frequency f_{bias} = 100 mHz. Note that the modulated signal shows a frequency of f_{mod} = 200 mHz, which is exactly twice the value of the bias frequency. This result suggests that the scattering is being somehow tuned by the magnetization of the sample, more precisely by

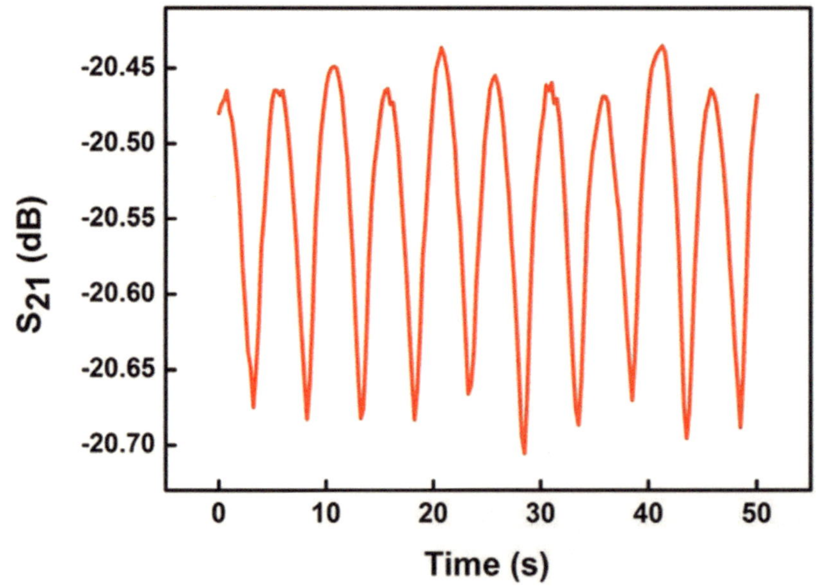

Figure 8. Experimental S_{21} parameter time evolution for the experiment depicted in **Figure 5** with $f_{antennas}$ = 1.2 GHz and f_{bias} = 100 mHz.

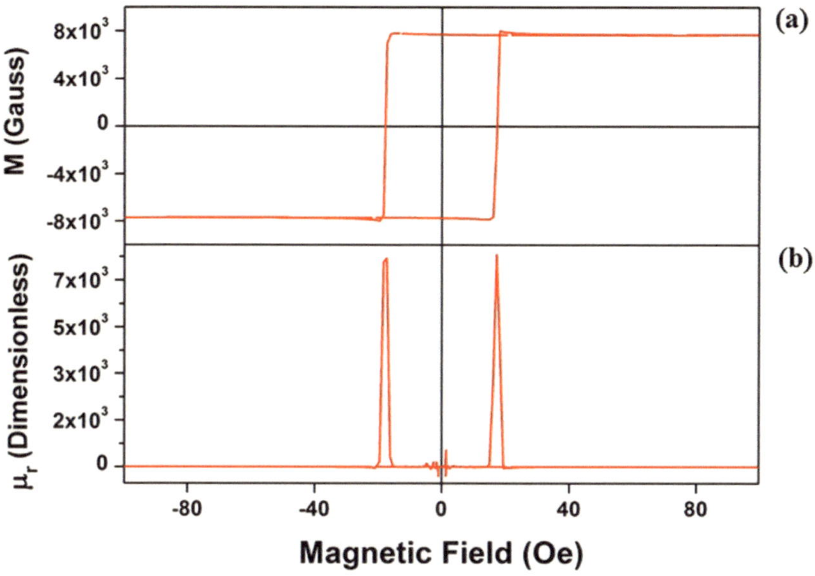

Figure 9. (a) Hysteresis loop of the $Fe_{2.25}Co_{72.75}Si_{10}B_{15}$ sample. (b) Relative magnetic permeability of the $Fe_{2.25}Co_{72.75}Si_{10}B_{15}$ sample.

its magnetic permeability μ_r. This doubled frequency is due to the symmetric shape of the microwire hysteresis loop ($|\mu_r(H_{bias})|) = |\mu_r(-H_{bias})|$, as can be seen in **Figure 9** in which the experimental hysteresis loop of the wire, measured using a quantum design vibrating sample magnetometer, as well as the low-frequency permeability, which reaches values of 8×10^3, has been plotted.

Moreover, the use of even a weak magnetostrictive microwire, as the one chosen, allows us to modify its magnetic permeability by application of mechanical stresses. **Figure 10** displays the result of an analogous experiment as the one depicted in **Figure 5** but with the addition of a tensile device. Such device allows the application of different mechanical stresses to the microwire during the experiment. It can be noticed how the modulated pattern of S21 is

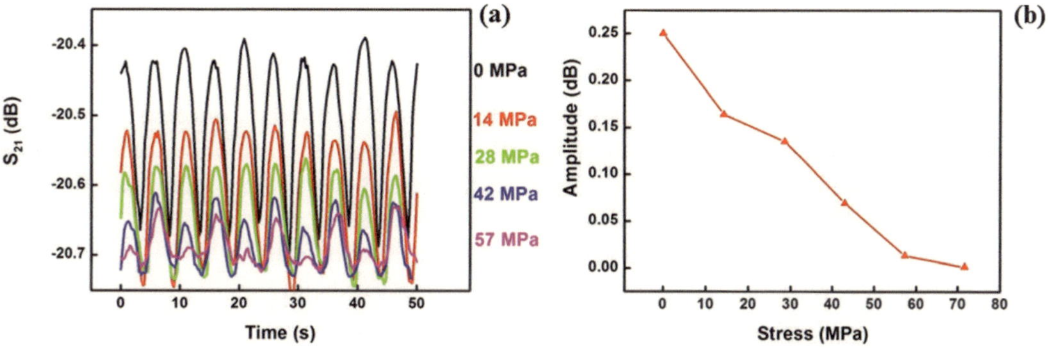

Figure 10. (a) Experimental S_{21} parameter time evolution for the $F_{2.25}Co_{72.75}Si_{10}B_{15}$ microwire subjected to different mechanical stresses. (b) Peak-to-peak amplitude of the S_{21} modulated signal as a function of the applied mass load (Figure published with copyright permission from Appl. Phys. Lett. 2014;105:092505(1)-092505(4)).

modified under the application of the different mechanical stresses. In this case, the application of the stress induces a magnetoelastic anisotropy, with strength given by the product of the stress and the magnetization constant, which decreases the relative permeability, μ_r, of the microwire.

The physical aspects underlying the experiments reported here are as follows. The modulation of S_{21} parameter observed in **Figures 2** and **4** describes the modulation of the reflected power P_2 at the receiving antenna position. The intensity of the wave reflected or scattered by the microwire shall depend on its impedance Z_2 as shown in Eq. (2). According to Eqs. (2) and (3), the bias field induces periodic changes of S_{21} as a consequence of the modulation that this field produces on μ_r as described by **Figure 8**. The periodic variation of μ_r gives rise through Eq. (2) to Z_2 and P_2 modulations that finally originates the experimental behavior summarized in **Figures 8** and **10**. The figure of merit, β, which we are interested in, is given by the following expression:

$$\beta = \frac{\Delta S_{21}}{S_{21}} = \alpha \Delta \mu_r \tag{4}$$

where ΔS_{21} is the change of S_{21} originated by a change in permeability of strength $\Delta\mu_r$. Accordingly to previous considerations, the parameter should be estimated through the dependence of both S_{21} on P_2 and from that of Z_2 on P_2.

The last one may be derived after using a standard method for calculating the scattering of electromagnetic waves by cylindrical ideal conductors that can be found elsewhere in the bibliography [40]. Such procedure is strictly based on solving the boundary conditions of the incident and reflected fields at the interphase r = a that in our case is the radius of the microwire metallic core. If we use the same geometry than in standard calculations [40] but considering a cylinder with finite conductivity and after expressing the magnetic and electric fields as linear combinations of Bessel, I_i, and Hankel, H_i, functions, their boundary conditions yield, for the transversal magnetic (TM) and transversal electric (TE) cases, reflection coefficients ($\widetilde{a_i}^{TE}$, $\widetilde{a_i}^{TM}$) of the Hankel, $H_i(k_1r)$, functions describing the outgoing scattered wave [38]. Contrary to the perfect conductor, reflection coefficients, $\widetilde{a_i}^{TE}$ and $\widetilde{a_i}^{TM}$, both function on Z_2 and k_2 which, in turn, are dependent on the magnetic permeability (see Eq. (2) and the following expression):

$$k_2 = \sqrt{i\omega\sigma\mu_0\mu_r} \tag{5}$$

Thus, it is possible to change $\widetilde{a_i}^{TE}$ and $\widetilde{a_i}^{TM}$ by tuning the magnetic state of the cylinder.

Some estimative considerations are outlined for a first-order approximation that corresponds to the case in which only a_0 coefficients are relevant. In order to establish the correlation between the experimental parameter S_{21} and the theoretical scattering coefficients $\widetilde{a_0}^{TE}$ and $\widetilde{a_0}^{TM}$, we formulate the S_{21} parameter as a function of the scattering coefficients $\widetilde{a_0}^{TE}$ and $\widetilde{a_0}^{TM}$, by expressing the input and output power in Eq. (4) as the average value of the Poynting vector in the positions corresponding to both antennas. The explicit dependence of S_{21} with the theoretical scattering parameters is given by the following expression:

$$S_{21} = 20log_{10}\left(\frac{\left(1 + \widetilde{a_0}^{circ}(\mu_r)\right)E^i_{in}(kr_2)}{E^i_{in}(kr_1)}\right)^2 \tag{6}$$

where E_{in}^{i} stands for the incident electric field; r_1 and r_2 represent the positions of the emitting and receiving antennas, respectively; and $\tilde{a}_0{}^{circ}$ stands for the scattering parameter for circular polarized radiation. e $\tilde{a}_0{}^{circ}$ can be estimated as the mean value of the TE and TM modes.

In summary, it has been experimentally shown that the microwave scattered intensity produced by a single microwire can be controlled by tuning its permeability. This permeability may be modified with the application of an external bias field, leading to non-negligible effects in the scattering. In addition, if the cylinder is magnetostrictive, the scattering is also sensitive to mechanical stresses. These experimental results are promising for developments in this field including in situ and in vivo biomedical magnetoelastic experiments, taking advantage of the biocompatible nature of the microwire Pyrex cover. So far, these kinds of measurements were only possible by using GMI or magnetoelastic resonance-based devices. However, the use of GHz frequency devices would allow the development of sensors operating at much longer distances and with a higher information transfer rate.

2.4. Liquid pressure wireless sensor based on magnetostrictive microwires for applications in cardiovascular localized diagnostics

The present work shows an application of such microwire as a wireless stress sensor with the particular application of pressure detection in a hydraulic circuit simulating cardiovascular conditions.

Two kinds of experiments have been performed. Both using a $Fe_{2.25}Co_{72.75}Si_{10}B_{15}$ amorphous magnetostrictive microwire with negative magnetostriction constant $\lambda_s = -0.9 \times 10^{-6}$ and saturation magnetization, $\mu_o M_s = 1.1$ T of metallic core radius of a = 33 μm, and total radius, including the Pyrex outer shell, of 50 μm. The coercive field measured under a 10 Hz applied field is 3 Oe. The relative permeability of the microwire changes between 800 and 2 when the applied field varies between 0 and 20 Oe, 30, respectively. We selected this composition because, as stated above, Co-rich microwires are well known to exhibit a noticeable high GMI effect [32].

The first experiment consists in tagging a silicon test tube (**Figure 11(a)**) with a single microwire, in order to show the possibility of using this kind of material as a contactless strain-sensing element. The second one has been designed to test the possibilities of this material to detect changes of the blood flow pumping both in an endovascular prosthesis and in an artery. With this second aim, either a piece of an artery or prosthesis (**Figure 11(b)**) sensorized by means of a ring of Co-based magnetostrictive microwire is placed in a hydraulic circuit connected to a pulsatile ventricular assist system (Abiomed/AB5000). A water seal and two resistors in parallel have been situated in order to control over fluid pressure. The hydraulic circuit described allows a fluid flow pumping frequency of 40/60 = 0.666 Hz, which is a signal with a period of 1.5 s corresponding to a pump of 40 beats/min. A solution of 0.33% agar has been used to make the fluid through the circuit to have a similar blood viscosity at 37°C. In order to register, instantaneously, the fluid pressure through the circuit of two probes, connected to a pressure detector, was invasively connected in distal and proximal positions, respectively, with respect to the sensing element (microwire ring). Fluid pressure can be calculated as a ratio between these two values.

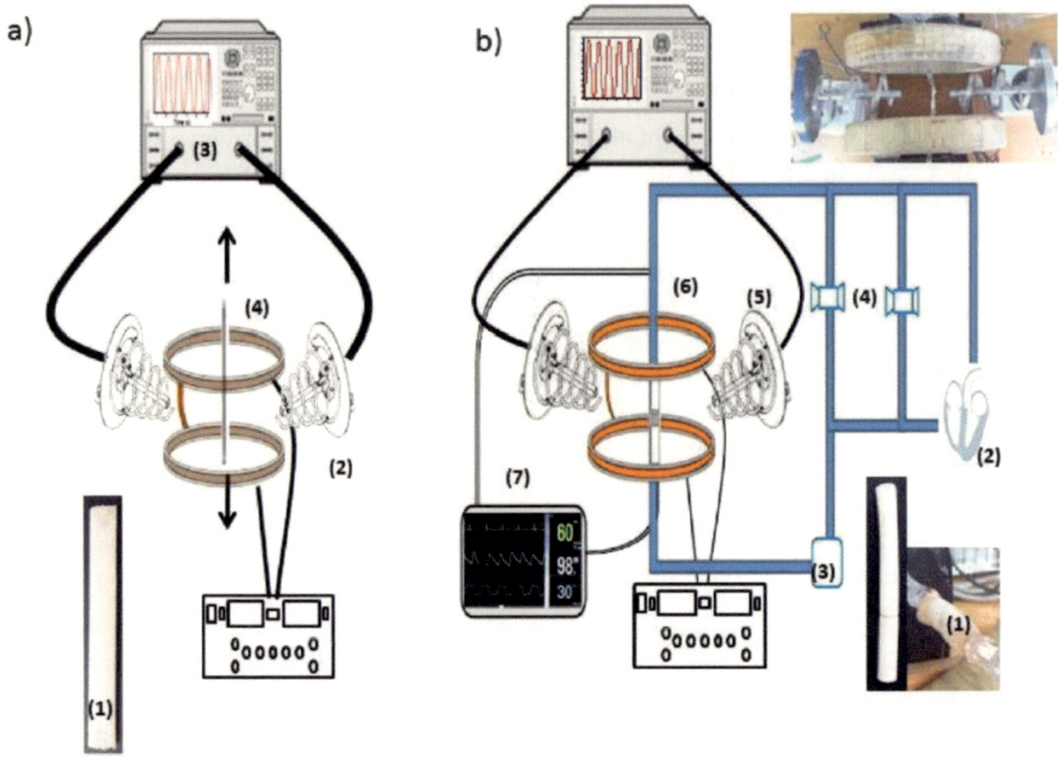

Figure 11. Experimental setups (a) on silicon test tube (1) under stress where emitting and receiving antennas (2) are connected to spectrum analyzer (3) and Helmholtz coils generate low-frequency magnetic field (4); (b) to detect changes of the blood flow pumping both in an endovascular prosthesis and in an artery. A piece of an artery or prosthesis (1) sensorized by means of a ring of Co-based magnetostrictive microwire placed in a hydraulic circuit connected to a pulsatile ventricular assist system (Abiomed/AB5000) (2), water seal (3), and two resistors in parallel (4) to control over fluid pressure, emitting and receiving antennas (5), Helmholtz coils (6), and invasive pressure detector (7) (Figure published with copyright permission from AIP ADVANCES5, 087132 (2015)).

Both specimens, sensorized silicon test tube (see **Figure 11a**) and endovascular prosthesis or artery connected to the hydraulic circuit (see **Figure 11b**), were placed between two helical antennas working at 1.2 GHz with circular polarized radiation. Both antennas were connected to a two-port Agilent E8362B Network Analyzer. The distance between antennas was large enough to ensure that a human body could be comfortably placed between them and that the far-field contribution dominates the near-field one. In addition to that, two Helmholtz coils with a customized electronic setup were used to apply an ac-bias magnetic field. Such experimental arrangement allows us to measure the scattering parameters of the system S_{21}.

Figure 12a displays the experimental result of measuring the S_{21} parameter over time for the aforementioned setup in the case of the first experiment with an emission frequency of the antennas set at $f_{antennas}$ = 1.2 GHz. The modulation of the S_{21} parameter shown in **Figure 12a** is produced by the application of a bias field with frequency f_{bias} = 100 mHz although the modulated signal shows a frequency of f_{mod} = 200 mHz, which is exactly twice the value of the bias frequency due to the tuning of the scattering by sample magnetization. Moreover, the

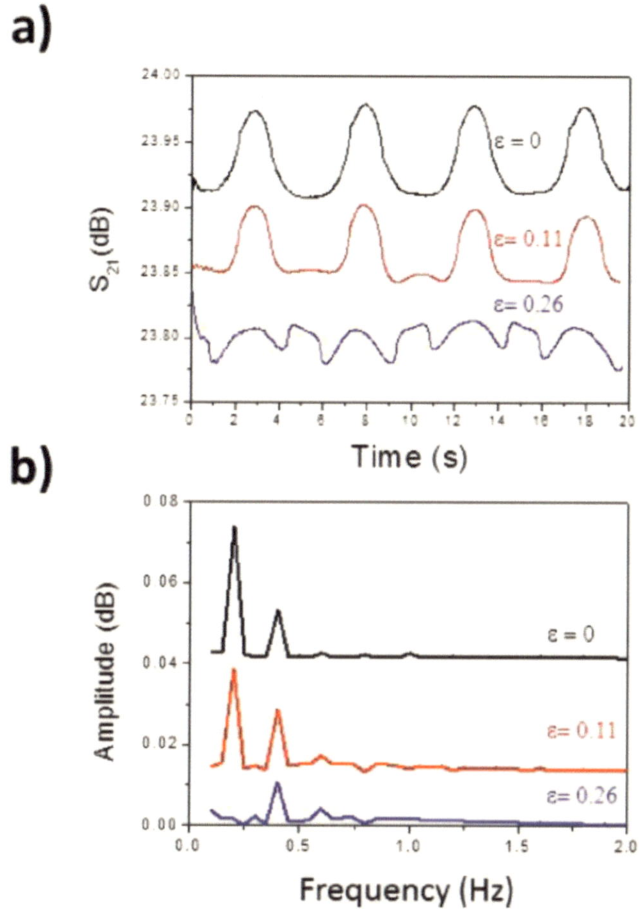

Figure 12. Influence of strain on silicon test tube on scattering attenuation coefficients, S_{21}, as a function of time (a) and corresponding fast Fourier transformation (FFT) curves (b).

use of even a weak magnetostrictive microwire, as the one chosen, allows us to modify its magnetic permeability by application of mechanical stresses. It can be noticed how the modulated pattern of S_{21} is modified under the application of the different mechanical stresses and correlated with the silicon test tube strain.

As shown in Ref. [38], the application of the stress induces a magnetoelastic anisotropy, with strength given by the product of the stress and the magnetization constant, which decreases the relative permeability, μ_r, of the microwire. **Figure 12b** shows the fast Fourier transform of curves in **Figure 12a** to quantify the frequency and amplitude of the signals. Two frequencies are observed, 200 and 400 mHz, respectively, with a clear amplitude decrease for the first one as a function of test tube strain, while the second one remains almost constant. From these data the relationship between the silicon strain and the reflectivity of the microwire can be roughly estimated. A reflectivity decrease of 0.033 dB is associated to a strain $\varepsilon = 0.26$.

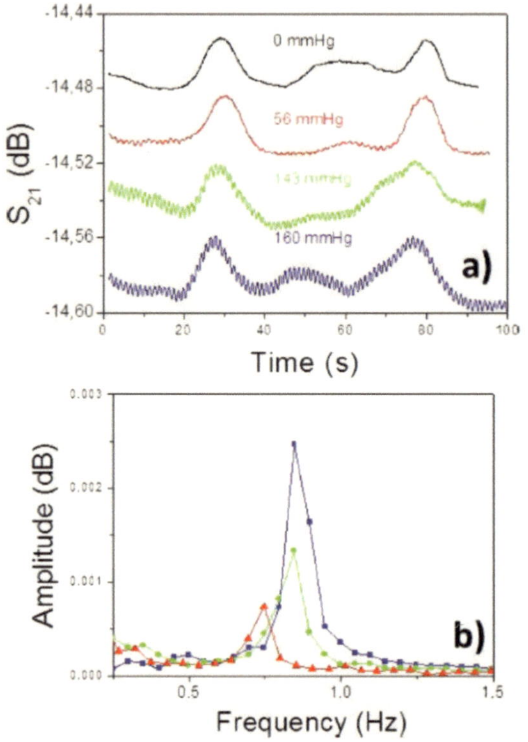

Figure 13. Experimental results of measuring the S_{21} parameter in the case of the sensorized arterial prosthesis using 1.2 GHz and a bias field frequency f_{bias} = 8 mHz as the function of fluid pressure (a) and corresponding fast Fourier transformation (FFT) (b) for 56 mmHg (▲), 143 mmHg (•), and 160 mmHg (■) (Figure published with copyright permission from AIP ADVANCES5, 087132 (2015)).

Figure 13a shows the experimental results of measuring the S_{21} parameter in the case of the sensorized arterial prosthesis using 1.2 GHz and a bias field frequency f_{bias} = 8 mHz as function of fluid pressure (from 0 to 160 mmHg). An initial modulated signal of 16 mHz is observed, but, as the pressure increases, a second superimposed signal with a frequency of 666 mHz can be seen. The amplitude of this second signal increases as fluid pressure grows (see the inset in **Figure 13**).

A crucial point in this experiment should be noted. This is the relationship between the frequencies of the bias field and the fluid flow pumping. Only a very low value of bias field frequency allows the observation of the effect. In this case the fast Fourier transformation (FFT) in **Figure 13b** shows two main peaks: the first one at 16 mHz with no sensible variation due to pressure and a second one at around 700 mHz strongly affected by pressure changes.

The same experiment has been performed in a bovine artery for fluid pressure between 100 and 235 mmHg. **Figure 14** shows pressure influence on S_{21} parameter as a function of time under the same conditions of the previous case of the prosthesis. The greater flexibility of the artery is clearly shown in the evolution of the fast Fourier transform of these signals with increasing system pressure.

http://dx.doi.org/10.5772/intechopen.70354

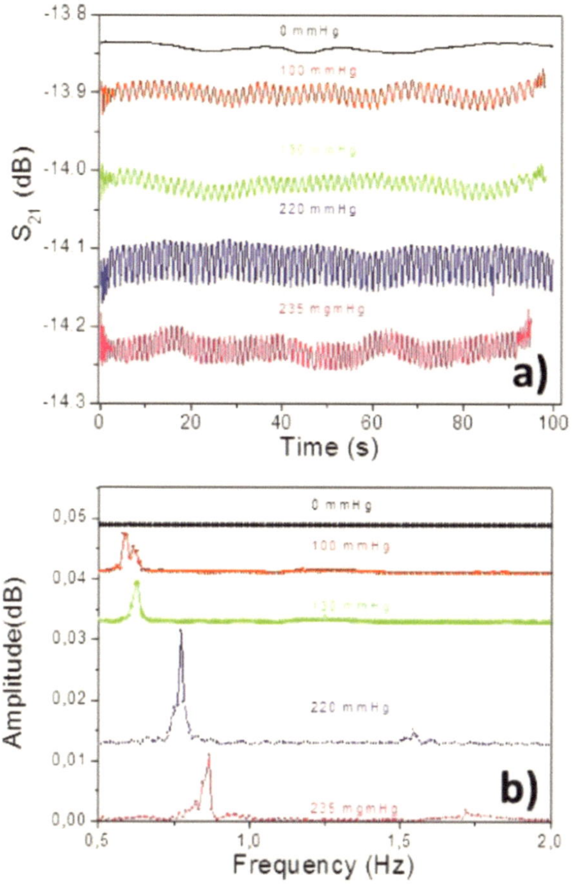

Figure 14. Experimental results of measuring the S_{21} parameter in the case of the sensorized bovine artery using 1.2 GHz and a bias field frequency f_{bias} = 8 mHz as function of fluid pressure (a) and corresponding fast Fourier transformation (FFT) (b) for 56 mmHg (▲), 143 mmHg (•), and 160 mmHg (■) (Figure published with copyright permission from AIP ADVANCES5, 087132 (2015)).

Figure 15 compares reflectivity versus pressure for both elements where the same changes in the fluid pressure involve higher increment in reflectivity for the artery in concordance with its higher elasticity.

In conclusion, we proposed the flexible magnetic element able to be integrated both in artery and prosthesis suitable for wireless localized blood pressure monitoring [22]. The sensor made of a ring of glass-covered magnetic microwire is simple and inexpensive and could be detected by means of a simple exciting and detecting set up able of emitting and detecting microwaves simultaneously applying a low-frequency magnetic field. The reflectivity of the microwire is determined by mechanical changes. In the experimental study, a piece of a cardiovascular prosthesis and a piece of an artery both sensed with a ring of magnetic microwire have been situated in a hydraulic setup simulating cardiovascular human circuit. Reflectivity changes of the sensor show its capability of measuring pressure variations in the circuit.

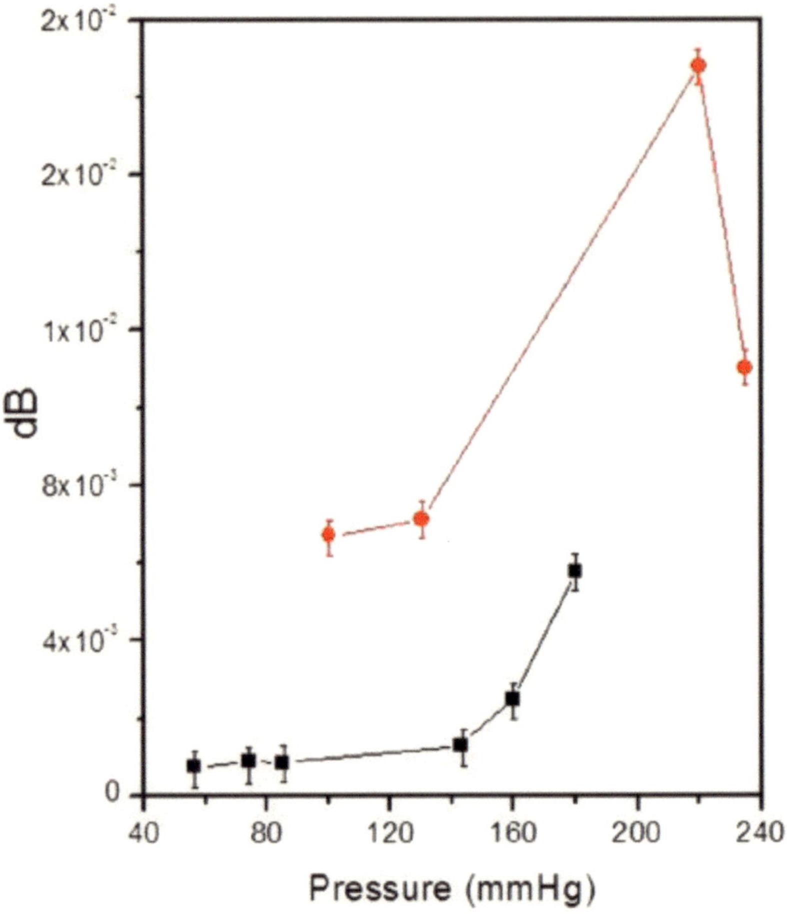

Figure 15. Reflectivity versus pressure for sensorized arterial prosthesis (■) and bovine artery (•) (Figure published with copyright permission from AIP ADVANCES5, 087132 (2015)).

3. Conclusions

In conclusion, an electromagnetic field induces an electric current along the microwire that can be modulated by the application of an external ac magnetic field, an effect known as magnetoimpedance. The maximum induced electric current, as well as the maximum ac modulation, occurs for frequencies determined by the microwire length. The modulation also varies as a function of the dc-applied field and takes a maximum when the dc field is close to the coercivity of the microwire. In fact, when the dc-field strength equals the coercive field, the variation of the permeability induced by the ac-superimposed field reaches a maximum value. This effect can be used for different sensing applications.

It is also interesting to note that for driving field frequencies corresponding to magnetoelastic or ferromagnetic resonance frequencies of microwires, small variations of the applied field or stress can give rise to strong changes in permeability. In order to increase the sensitivity of the system, it is also important to select the frequency in that range for which the couple of antennas is close to resonance in wires.

It must be noticed that we had outlined an estimation of the wire sensitivity without considering the tensor form of the permeability at the microwave range. A more rigorous analysis could be carried out, but to account for the order of magnitude of the experimental results reported in this work, the considered approximation seems to be sufficiently illustrative.

In summary, it has been experimentally shown that the microwave scattered intensity produced by a single microwire can be controlled by tuning its permeability. This permeability may be modified with the application of an external bias field, leading to non-negligible effects in the scattering. In addition, if the cylinder is magnetostrictive, the scattering is also sensitive to mechanical stresses. These experimental results are promising for future developments in this field including in situ and in vivo biomedical magnetoelastic experiments, taking advantage of the biocompatible nature of the microwire Pyrex cover.

Acknowledgements

The author wants to acknowledge the Spanish Ministry of Economy and Competitiveness for its support via the projects MAT2013-49847-EXP and MAT2015-67557-C2-1-P and Comunidad de Madrid for support S2013/MIT-2850 NANOFRONTMAG-CM.

Author details

Pilar Marín

Address all correspondence to: mpmarin@fis.ucm.es

Departamento de Física de Materiales, Instituto de Magnetismo Aplicado, Universidad Complutense de Madrid, Spain

References

[1] Marín P, López M, Agudo P, Vázquez M, Hernando A. Applications of amorphous samples presenting high magnetomechanical coupling during the first stages of nanocrystallisation process. Sensors and Actuators A: Physical. 2001;**91**(1-2):218-222. DOI: https://doi.org/10.1016/S0924-4247(01)00464-2

[2] Vázquez M. Handbook of Magnetism and Advanced Magnetic Materials. In: Kronmuller H, Parkin S, editors. United Kingdom: Wiley, Chichester; 2007. pp. 2193-2226

[3] Chiriac H, Tibu M, Moga AC, Herea DD. Magnetic GMI sensor for detection of biomolecules. Journal of Magnetism and Magnetic Materials. 2005;**293**(1):671-676. DOI: https://doi.org/10.1016/j.jmmm.2005.02.043

[4] Mohri K, Honkura Y. Amorphous wire and CMOS IC based magneto-impedance sensors —Origin, topics, and future. Sensor Letters. 2007;**5**(1):267-270(4). DOI: https://doi.org/10.1166/sl.2007.082

[5] Kaniusas E, Mehnen L, Pfützner H. Magnetostrictive amorphous bilayers and trilayers for thermal sensors. Journal of Magnetism and Magnetic Materials. 2003;**254-255**:624-626. DOI: http://doi.org/10.1016/S0304-8853(02)00922-8

[6] Taylor GF. A method of drawing metallic filaments and a discussion of their properties and uses. Physical Review B. 1924;**23**:655-658. DOI: https://doi.org/10.1103/PhysRev.23.655

[7] Baranov SA. Evaluation of the distribution of residual stresses in the cord of amorphous microwire. Metal Science and Heat Treatment. 2001;**43**(3):167-168. DOI: 10.1023/A:1010569906136

[8] Vázquez M, Zhukov AP, Aragoneses P, Arcas J, Garcia-Beneytez JM, Marin P, Hernando A. Magneto-impedance in glass-coated CoMnSiB amorphous microwires. IEEE Transactions on Magnetics. 1998;**34**(3):724-728. DOI: 10.1109/20.668076

[9] Vázquez M. Magnetic bistability of amorphous wires and sensor applications. IEEE Transactions on Magnetics. 1994;**30**(2):907-912. DOI: 10.1109/20.312442

[10] Marín P, Vázquez M, Arcas J, Hernando A. Thermal dependence of magnetic properties in nanocrystalline FeSiBCuNb wires and microwires. Journal of Magnetism and Magnetic Materials. 1999;**203**(1-3):6-11. DOI: http://doi.org/10.1016/S0304-8853(99)00173-0

[11] Vázquez M, Marin P, Arcas J, Hernando A, Zhukov AP, González J. Influence of nanocrystalline structure on the magnetic properties of wires and microwires. Textures and Microstructures. 1999;**32**(1-4):245-267. DOI: dx.doi.org/10.1155/TSM.32.245

[12] Beach RS, Berkowitz AE. Giant magnetic field dependent impedance of amorphous FeCoSiB wire. Applied Physics Letters. 1994;**64**:3652. DOI: http://dx.doi.org/10.1063/1.111170

[13] Vazquez M, Chen DX. The magnetization reversal process in amorphous wires. IEEE Transactions on Magnetics. 1995;**31**(2):1229-1238. DOI: 10.1109/20.364813

[14] Kraus L, Infante G, Frait Z, Vázquez M. Ferromagnetic resonance in microwires and nanowires. Physical Review B. 2011;**83**:174438 (1)-174438 (11). DOI: https://doi.org/10.1103/PhysRevB.83.174438

[15] Marín P, Marcos M, Hernando A. High magnetomechanical coupling on magnetic microwire for sensors with biological applications. Applied Physics Letters. 2010;**96**:262512(1)-262512(3). DOI: http://dx.doi.org/10.1063/1.3459140

[16] Herrero-Gómez C, Marín P, Hernando A. Bias free magnetomechanical coupling on magnetic microwires for sensing applications. Applied Physics Letters. 2013;**103**:142414(1)-142414(4). DOI: http://dx.doi.org/10.1063/1.4821777

[17] Marín P, Cortina D, Hernando A. High-frequency behavior of amorphous microwires and its applications. Journal of Magnetism and Magnetic Materials. 2005;**290-291**(2):1597-1600. DOI: https://doi.org/10.1016/j.jmmm.2004.11.255

[18] Makhnovskiy DP, Panina LV, Garc_ıa C, Zhukov A, González J. Experimental demonstration of tunable scattering spectra at microwave frequencies in composite media containing CoFeCrSiB glass-coated amorphous ferromagnetic wires and comparison with theory. Physical Review B. 2006;**74**:064205(1)-064205(11). DOI: 10.1103/PhysRevB.74.064205

[19] Luo Y, Peng HX, Qin FX, Ipatov M, Zhukova V, Zhukov A, Gonzalez J. Fe-based ferromagnetic microwires enabled meta-composites. Applied Physics Letters. 2013;**103**:251902. DOI: http://dx.doi.org/10.1063/1.4850196

[20] Marín P, Cortina D, Hernando A. Electromagnetic wave absorbing material based on magnetic microwires. IEEE Transactions on Magnetics. 2008;**44**(11):3934-3937. DOI: 10.1109/TMAG.2008.2002472

[21] Herrero-Gómez C, Aragón AM, Hernando-Rydings M, Marín P, Hernando A. Stress and field contactless sensor based on the scattering of electromagnetic waves. Applied Physics Letters. 2014;**105**:092405(1)-092405(4). DOI: http://dx.doi.org/10.1063/1.4894732

[22] Aragón AM, Hernando-Rydings M, Hernando A, Marín P. Liquid pressure wireless sensor based on magnetostrictive microwires for applications in cardiovascular localized diagnostic. AIP Advances. 2015;**5**:087132(1)-087132(7). DOI: 10.1063/1.4928605

[23] Hao Y, Foster R. Wireless body sensor networks for health-monitoring. Physical Meas.2008;**29**(11):R27-R56. DOI: 10.1088/0967-3334/29/11/R0

[24] Hernando Rydings M, Marín Palacios P, Aragón-Sánchez AM, Bravo Ruiz E, López-Domínguez V, Martínez López I, Fernández Pérez C, Bilbao González A, Javier Serrano Hernando F, Vega Manrique R, Hernando Grande A. Development of a telemetric system for postoperative follow-up of. Journal of the American Heart Association. 2016;**5**(e003608):1-12. DOI: 10.1161/JAHA.116.003608

[25] Taylor GF. A method for drawing metallic filaments and discussion of their properties and uses. Physical Review B. 1924;**23**:6555-6560. DOI: https://doi.org/10.1103/PhysRev.23.655

[26] Baranov AS, Larin VS, Torkunov AV, Zhukov AP, Vázquez M. Magnetic properties of glass insulated amorphous microwires. In: Vázquez M, Hernando A, editors. Nanocrystalline and Non-crystalline Materials. Singapore: World Scientific; 1995. pp. 567-571

[27] Gorriti AG, Marín P, Hernando A. Microwave power absorption by microwires under tensile stress. Sensor Letter. 2004;**7**(3):1-4. DOI: https://doi.org/10.1166/sl.2009.1058

[28] King RWP, Wu TT. The imperfectly conducting cylindrical transmitting antenna. IEEE Transactions on Antennas and Propagation. 1966;**14**(5):524-534. DOI: 10.1109/TAP.1966.1138733

[29] Makhnovskiy DP, Panina LV, Mapps DJ. Field-dependent surface impedance tensor in amorphous wires with two types of magnetic anisotropy: Helical and circumferential.

Physical Review B. 2002;**64**(14): 144424-144441. DOI: https://doi.org/10.1103/PhysRevB.63.144424

[30] Usov NA, Antonov AS, Lagarkov AN. Theory of giant magneto-impedance effect in amorphous wires with different types of magnetic anisotropy. Journal of Magnetism and Magnetic Materials. 1998;**185**:159-173. DOI: 10.1016/S0304-8853(97)01148-7

[31] Sandacci SI, Makhnovskiy DP, Panina LV. Valve-like behavior of the magnetoimpedance in the GHz range. Journal of Magnetism and Magnetic Materials. 2004;**272**:1855-1857. DOI: 10.1016/j.jmmm.2003.12.829

[32] Lofland SE, Bhagat SM, Dominguez M, García-Beneytez JM. Low-field microwave magnetoimpedance in amorphous microwires. Journal of Applied Physics. 1999;**85**(8): 4442-4444. DOI: http://dx.doi.org/10.1063/1.370368

[33] Makhnovskiy DP, Panina LV. Field dependent permittivity of composite materials containing ferromagnetic wires. Journal of Applied Physics. 2003;**93**:4120. DOI: http://dx.doi.org/10.1063/1.1557780

[34] Makhnovskiy DP, Panina LV, Garcia C, Zhukov AP, Gonzalez J. Experimental demonstration of tunable scattering spectra at microwave frequencies in composite media containing CoFeCrSiB glass-coated amorphous ferromagnetic wires and comparison with theory. Physical Review B. 2006;**74**:064205. DOI: https://doi.org/10.1103/PhysRevB.74.064205

[35] Hanson G. Fundamental transmitting properties of carbon nanotube antennas. IEEE Transactions on Antennas and Propagations. 2005;**53**(11):3426-3435. DOI: 10.1109/TAP.2005.858865

[36] Panina LV, Makhnovskiy DP, Morchenko AT. Tunable permeability of magnetic wires at microwaves. Journal of Magnetism and Magnetic Materials. 2015;**383**:120-125. DOI: 10.1016/j.jmmm.2014.11.051

[37] Hernando A, López-Domínguez V, Ricciardi E, Osiak K, Marín P. Tuned scattering of electromagnetic waves by a finite length ferromagnetic microwire. IEEE Transactions on Antennas and Propagations. 2016;**64**(3):1112-1115. DOI: 10.1109/TAP.2015.2513428

[38] Herrero-Gómez C, Aragón AM, Hernando-Rydings M, Marín P, Hernando A. Stress and field contactless sensor based on the scattering of electromagnetic waves by a single ferromagnetic microwire. Applied Physics Letters. 2014;**105**:092505(1)-092505(4). DOI: http://dx.doi.org/10.1063/1.4894732

[39] Hernando A, Vázquez M, Barandiarán JM. Metallic glasses and sensing applications. Journal of Physics E: Scientific Instruments. 1988;**21**(12):1129. DOI: https://doi.org/10.1088/0022-3735/21/12/002

[40] Jackson JD. Classical Electrodynamics. 3rd ed. New York: EEUU, Wiley; 1998

Chapter 3

Thin-Film Magneto-Impedance Sensors

Alfredo García-Arribas, Eduardo Fernández and
David de Cos

Additional information is available at the end of the chapter

http://dx.doi.org/10.5772/intechopen.70084

Abstract

We review the state-of-the-art of thin films displaying the magneto-impedance (MI) effect, with focus on the aspects that are relevant to the successful design and operation of thin film-based magnetic sensors. After a brief introduction of the MI effect, the materials and geometries that maximize the performance, together with the measurement procedure for their characterization, are exposed. A nonexhaustive survey of applications is included, mostly with the aim of displaying the capabilities of thin film structures in the field of magnetic sensing, and the variety of topics covered by them. A special emphasis is made on some concepts that are not commonly treated in the literature, such as the influence of the measuring circuit on the magneto-impedance ratio, the geometry optimization by means of numerical simulation by finite element methods, or noise measurements on thin films. Additionally, a brief description of the patterning procedure by photolithography is included, since the major advantage of thin film sensors over other types of magneto-impedance materials as ribbons or wires is the possibility of patterning the sensible element in micrometric shapes, and most of all, their easy integration with the interface microelectronic circuitry.

Keywords: magneto-impedance, thin film, magnetic sensor, photolithography, finite element methods

1. Introduction

The giant magneto-impedance (MI) effect accounts for the large change of the electrical impedance experienced by a soft magnetic material when subjected to an external magnetic field. It was proposed as a very sensitive method to detect small magnetic fields during the 1990s [1, 2]. However, the basic underlying principle is a completely classical electromagnetic effect that was already studied by Lord Rayleigh in 1887 [3] and with much greater detail by Kittel in 1946 [4]. It is based on the so-called skin effect, which accounts for the limited penetration on conductive materials of the electromagnetic field associated with an alternating current flowing through them. The exponential decrease of the amplitude of the fields from the surface of the conductor is characterized by the penetration depth δ given by

$$\delta = 1/\sqrt{\pi f \sigma \mu}, \tag{1}$$

where f is the frequency of the alternating current and σ and μ are the conductivity and the magnetic permeability of the material, respectively. The effective cross-section of the sample for the flow of the alternating current depends on the magnitude of the penetration depth. Therefore, the skin effect determines the impedance of the specimen, and if the sample is magnetic, the external magnetic field modifies the permeability and produces the magneto-impedance effect. In summary, the basics of MI are described by Maxwell equations and explained in classical treatises on electromagnetism [5, 6].

If the geometry of the sample is simple enough, the Maxwell equations can be solved analytically to obtain the impedance. In the case of a homogenous sheet of thickness $2a$ and infinite width and length, the result is [7]

$$Z = R_{dc} \sqrt{j}\theta \coth \sqrt{j}\theta \tag{2}$$

being R_{dc} the dc resistance, $j = \sqrt{-1}$ (the imaginary unit), and $\theta = a\sqrt{2\pi f \sigma \mu} = \sqrt{2}\, a/\delta$. A similar expression involving Bessel equations can be obtained for wires [7].

It is to be noted that the permeability involved in expressions (1) and (2) is the one responding to the alternating magnetic field created by the flowing current (**Figure 1**), which is transversal to the direction of current flow and is usually denominated transverse permeability.

The magneto-impedance is typically quantified as a difference ratio with respect to a reference impedance value as

$$\mathrm{MI}\ (\%) = \frac{Z - Z_{low}}{Z_{low}} \times 100 \tag{3}$$

where Z_{low} is almost always taken as the lowest measured value of the impedance. Habitually, Z_{low} corresponds to the impedance of the sample when it is magnetically saturated and its relative permeability is close to 1: $Z_{low} = Z_{sat} = Z(\mu_0)$. Experimentally, Z_{low} usually corresponds to the impedance measured at the maximum applied magnetic field $Z_{low} = Z\,(H_{max})$, which is supposed to magnetically saturate the sample. It can also be taken as the impedance at zero field $Z_{low} = Z(H = 0)$, if the MI curve is double-peaked (**Figure 1**).

One of the figures of merit for MI is the maximum value of the magneto-impedance given by Eq. (3) MI_{max}, which is reached when the value of Z is maximum. For a given frequency, and according to Eq. (1), this occurs when the permeability is maximum and the penetration depth is minimum. Therefore, the magnetization processes of the sample, which determine the magnitude of the transverse permeability, are paramount in achieving large MI_{max} ratios.

The second figure of merit is the maximum sensitivity s_{max}, which is actually the most important for applications. It expresses the maximum change of impedance experienced by the

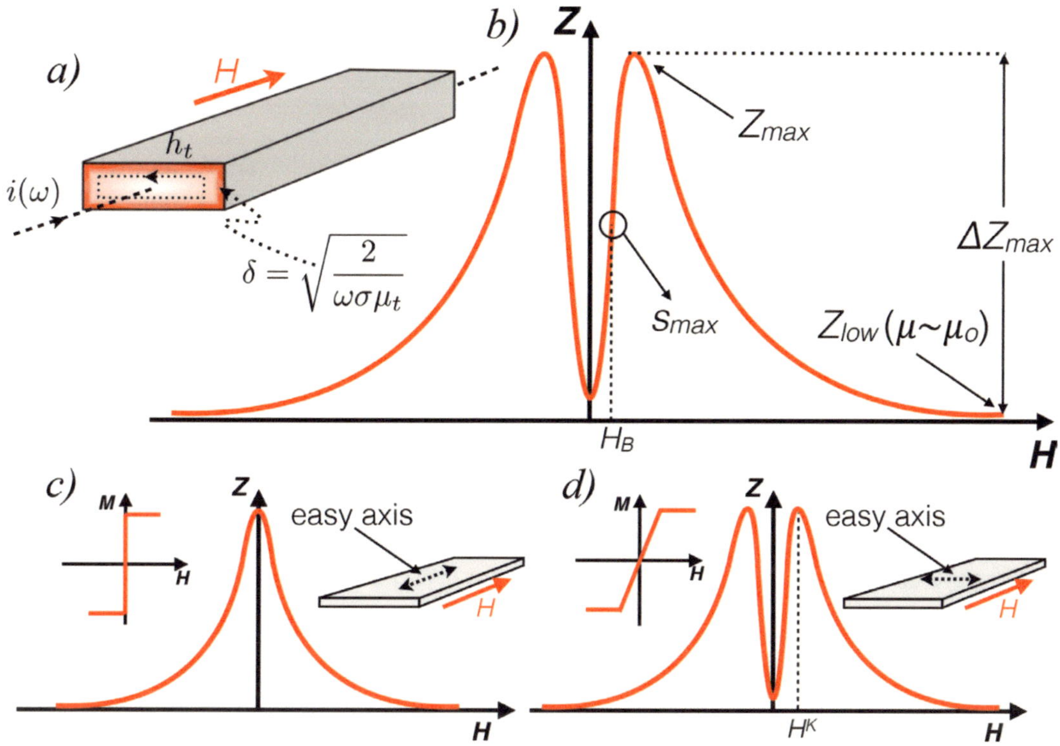

Figure 1. (a) Scheme of the magneto-impedance effect in a planar sample: the alternating current i accumulates on the surface due to the skin effect, whereas its magnetic field h_t excites the material according to its transverse permeability μ_t, which is modulated by the externally applied magnetic field H. (b) Typical magneto-impedance curve: the maximum impedance occurs when μ_t is largest, and the minimum when it is magnetically saturated. The maximum sensitivity s_{max} takes place at the point of the largest slope of the $Z(H)$ curve. The sample must be biased with a field H_B to obtain the maximum field sensitivity. (c) Single-peaked MI curve when the magnetic field H is applied in the direction of the magnetic easy axis (longitudinal anisotropy). (d) Double-peaked MI curve in the case of transverse anisotropy.

sample when the magnetic field acting on it varies, that is, the maximum slope of the $Z(H)$ curve. It also depends on the magnetization process. The most usual cases in planar samples are schematized in **Figure 1(c)** and **(d)**. In a sample with longitudinal anisotropy, the magnetization easy axis lays in the direction of the applied field and the flow of current. The maximum transverse permeability occurs at $H = 0$, and the $Z(H)$ curve presents a single peak. If the sample presents transverse anisotropy, the maximum transverse permeability takes place at the anisotropy field H^k, defined as [8]

$$H^k = \frac{2K}{\mu_0 M_s}, \tag{4}$$

where K is the anisotropy constant and M_s the saturation magnetization. This is the preferred case, since the curve $Z(H)$ presents two peaks and can display a great sensitivity if MI_{max} is large and H^k is small. In thin film MI sensors, it is extremely important to obtain a well-defined transverse magnetic anisotropy.

2. Thin film MI samples

2.1. Sample geometry

The best samples for large MI effect are amorphous wires and glass-coated microwires with very small negative magnetostriction [9, 10], especially those in which the configuration of the magnetic domains favors a large circular (transverse) permeability [11]. Values up to MI_{max} = 500% are reported in the best homogenous wires. MI has been also actively studied in amorphous ribbons [12, 13]. In both cases (wires and ribbons), the MI takes large values at frequencies of tens or hundreds of kHz, when the penetration depth equals the thickness of the sample, in the order of tens of microns. MI in thin films, however, requires much larger frequencies, typically of the order of 1 GHz for films of about 1 μm thick. This situation has two important consequences that will be discussed in the following sections: first, the measuring circuits for the characterization of the MI response must be adapted for such elevated frequencies, and second, from the point of view of the properties of the thin film material, adequate strategies must be adopted to increase the thickness of the sample without deteriorating the soft magnetic behavior.

To increase the magnitude of the MI effect in thin film materials, a sandwiched structure with a nonmagnetic conductor between two magnetic layers was early proposed [14–16], in which the magneto-inductive effect, created by the large permeability of the magnetic layers, allows to override the pure skin effect and achieve large impedance variations at moderate frequencies. The advantages of this layered structure were also incorporated in wire-shaped specimens to achieve MI ratios of 800%, the largest reported in the literature [17]. Analytical expressions similar to Eq. (2), but more complicated, can be obtained for some sandwiched geometries by imposing the suitable boundary conditions at the interface of the materials [18]. In sandwiched structures it is possible to devise two main configurations as schematized in **Figure 2**. If the outer magnetic layers completely enclose the central conductor, the layout resembles that of a core-shield wire, in which the magnetic flux created by the current encounters a closed path with high permeability. In contrast, the open-flux configuration, in which the magnetic and central layers have the same width, makes the flux created by the current

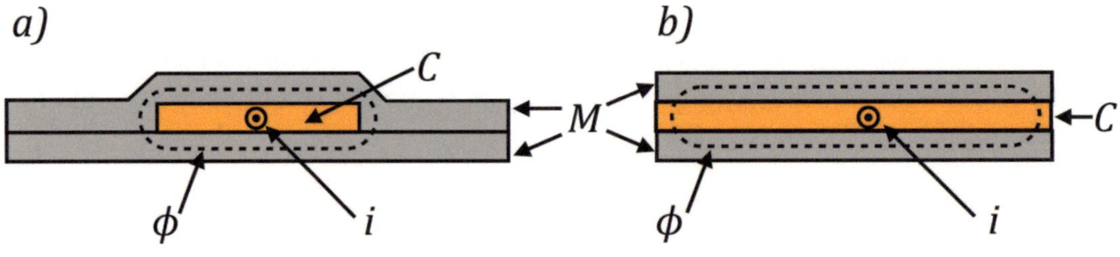

Figure 2. The sandwich structure *M/C/M* (*M* = magnetic layer, *C* = conductive, nonmagnetic layer) enhances the magneto-inductive response of thin film samples, overriding the poor influence of the skin effect at moderate frequencies (in the range of MHz). The figure shows the cross-section of two possible configurations for sandwiched structures. In (a) the magnetic layers enclose the central conductor, providing a high permeability closed path for the magnetic flux φ generated by the alternating current i (flowing perpendicular to the page). In (b), the flux lines must close by a nonmagnetic medium, reducing the magneto-inductive effect, but this structure is much easier to fabricate.

cross a low permeability path. The flux leakage worsens the MI behavior [19], but the open-flux configuration can be fabricated in a single step, which makes it preferable over the closed-flux one. Besides, the direct comparison between both configurations demonstrates that the MI performance gain found in closed-flux is not significant due to additional problems arising from the increased thickness of the magnetic layers in the edges of the structure [20].

Thin film MI materials are commonly patterned to adjust their lateral dimensions to the desired shape. The usual geometry is in the form of rectangular stripes, but other shapes as meanders or ellipsoids are used depending on the application. The patterning can be performed during the deposition or using photolithography techniques. More details will be given in Section 2.3.

2.2. Thin film MI materials

The most successful MI materials, in the form of wires and ribbons, are made from amorphous alloys (Fe-Co-Si-B or similar), rapid quenched from the melt. These materials, obtained by other methods, have also been used for thin film MI structures. In fact, values of MI_{max} close to 700% were early reported from a closed-flux sandwiched structure of CoSiB layers [21]. Such extraordinary performance was obtained due to the large thickness of the magnetic layers (2 μm) and the use of isolation SiO_2 layers between the central conductor and the outer magnetic layers. Also, amorphous FeSiB [22], CoNbZr [23], and FeCuNbSiB [24] films have been investigated in MI structures. But the dominant tendency in MI thin film structures is to use Permalloy (Py) as magnetic layer. With a nominal composition of $Fe_{20}Ni_{80}$, Permalloy is a very soft magnetic alloy with large permeability, no crystalline anisotropy, and nearly vanishing magnetostriction, which are key properties for great MI performance. Besides, the constituents of Py are abundant and cheap and its use in MI devices benefits from the intense research performed toward enhancing anisotropic magnetoresistance (AMR) sensors that use Py as base material [25].

Permalloy in the form of thin film is obtained mainly by vapor deposition methods. The films with a small thickness are satisfactorily soft, but the growth of thicker films is affected by the development of a columnar structure that defines an easy magnetization axis in the direction perpendicular to the plane of the sample. In this situation, the sample enters what is called a transcritical state [26], observed also in amorphous films [27], which ruins the magnetic softness and results severely detrimental for MI performance, since MI requires both magnetic softness and increased sample thickness. Therefore, it is very important to determine the optimum preparation conditions, as well as the limiting Py single-layer thickness that can be reached before entering the transcritical state. In low pressure sputtering deposition, this thickness is determined to be about 170 nm [28]. To increase the total effective thickness, it has been proposed to use thin spacers between successive Py layers, creating a thick magnetic multilayer in which all the Py films are below the critical thickness [29, 30]. Different materials are used as spacers (Ti, Cu, Ag, etc.). Their nature and optimum thickness has been investigated [31] and the usefulness of this approach proven [32]. **Figure 3** displays the typical layer configuration of a high performance MI structure, which combines the sandwich layout to take advantage for the magneto-inductive effect, with outer magnetic parts composed of a multilayer Py/Ti structure in which the Py layers are below the critical thickness.

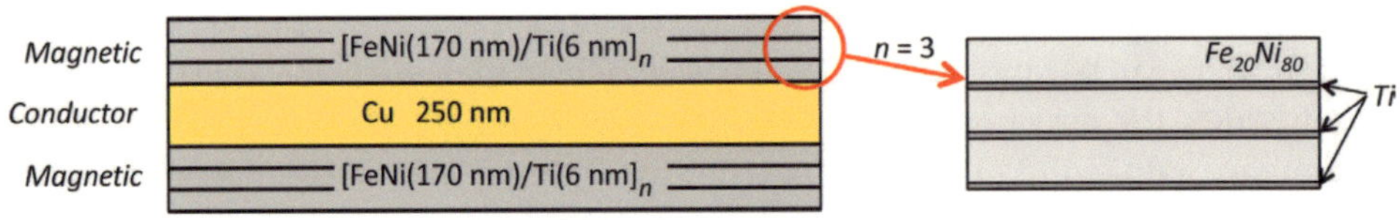

Figure 3. Scheme of the cross-section of a multilayered sandwiched structure with open-flux configuration: the outer magnetic layers are composed of Permalloy ($Fe_{20}Ni_{80}$) films below the transcritical thickness separated by thin titanium spacers. The central conductive layer is nonmagnetic.

Besides the sensitive element, other materials can play a role in MI devices. The obvious ones are metals for electric contacts. More exceptional are materials devised to produce a biasing effect on the MI element. As observed in **Figure 1**, the largest sensitivity obtained with a small bias field H_B is applied to the sample. In laboratory experiments, it is provided by an external coil; but for a functional device, biasing by an integrated thin film permanent magnet [33] or by the polarizing effect of the exchange bias effect has been demonstrated [34–36].

2.3. Fabrication of MI structures

Thin film MI structures are most commonly fabricated by physical vapor deposition methods, although electrodeposition [37] or flame plating [38] has also been used. Sputtering is preferred over evaporation. The process parameters (RF or DC sputtering, Ar pressure, power, deposition time, etc.) depend on the equipment used and the nature of the desired material. Glass or silicon wafers are used as substrates, although polymers are increasingly being used with the onset of flexible electronics [39–42]. As large sensitivity relies on a well-defined transverse anisotropy, an in-plane magnetic field is usually applied during deposition. The deposition process creates a continuous film on the substrate. For measurements and applications, the substrate can be cut into the desired shape, usually rectangular. Alternatively, metallic masks with perforated motives (rectangular slits, for example) can be used over the substrate during the deposition. However, thin film technology is excellently adapted to the patterning procedures of the microelectronic industry. Photolithography allows creating arbitrary shapes with extraordinary lateral resolution and very well-defined borders. Basically, photolithography transfers the desired pattern defined as opaque or transparent motives on a mask into a photosensitive resist deposited on the substrate. The resist is exposed to ultraviolet (UV) light (365 nm from Hg lamps in laboratory systems) through the mask and is sensitized in the illuminated regions, changing its resistance to a given chemical (developer). Therefore, upon developing, the nonresistant parts of the resist are removed, leaving the desired pattern on the substrate. The patterned resist can be used to shape the thin film in two different ways [43], exemplified in **Figure 4**. In *etching* processes, the material is deposited as a continuous film on the substrate and the resist is patterned on top of it afterwards. The material not protected by the resist is removed using a chemical or plasma etching. In a *lift-off* process, the resist is patterned directly on the substrate and the material is deposited afterwards on top of it. When the resist is dissolved (in acetone or similar), the material deposited on top of it is peeled off, leaving the desired pattern on the substrate. Lift-off requires an inwards profile in the patterned resist in order to produce well-defined, sharp edges, as illustrated in **Figure 4(b)**.

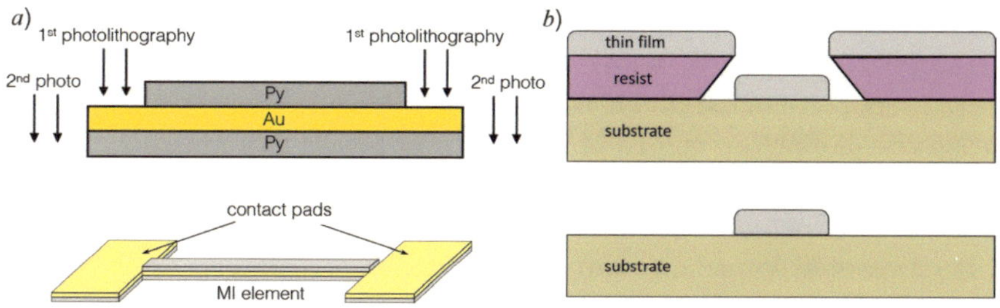

Figure 4. Example of photolithography processes to pattern thin film MI elements. (a) Photolithography is used to define the areas of the continuous film to be etched. The figure schematizes the two-step process to create the displayed structure. The first lithography defines the form of the element. The second reveals the central Au layer in the contact regions. Etching is performed by ion-milling in this case. (b) Illustration of the lift-off process in which the photoresist is patterned prior to the deposition of the thin film material. The inward profile of the resists facilitates the creation of sharp borders in the patterned material when the resist is dissolved, eliminating the material deposited on top of it.

Photolithography can be used to pattern MI samples in any desired shape. Although rectangles are most usual, meanders [37, 44] or ellipsoids [45] are also reported. Several photolithography steps can be performed sequentially to add pads for electric contacts, produce more complicated systems as closed-flux structures, or to integrate the sample in a functional device with the conditioning electronics. For example, successive photolithography steps are done to integrate a MI sample into a coplanar transmission line [45, 46].

The search for the optimum configuration of MI structures implies evaluating the impact on the MI performance of many fabrication parameters. For instance, in Py-based sandwiched structures, the effect of open- or closed-flux configuration [20], the thickness of the magnetic layers [47], the nature of the spacers between the magnetic layers [31], and the different thickness of the central conductor [48] have been experimentally assessed. The lateral dimensions of the samples play a fundamental role in the performance as well. While maximum miniaturization is beneficial for the development of MI-based microsensors, downsizing has large influence on the magnetic properties and general MI performance. For instance, the shape anisotropy in elongated samples competes with the transverse anisotropy induced during the preparation, modulating the sensitivity of the sample [49], and producing an inhomogeneous behavior over the sample that can be compensated by shaping the sample in the form of ellipsoids [45] or limiting the usable region of the device [50]. Reducing the lateral size increases the flux leakage in open-flux sandwiched structures [51]. Besides, the reduced size modifies the magnetic domain structure, since the influence of closure domains becomes increasingly important as the width of the sample is reduced [52]. At the present moment, the MI performance diminishes considerably in identical samples upon size reduction [53].

3. MI measurements in thin films

3.1. Measuring impedance at high frequency

Sandwiched thin films materials display large MI variations in the frequency range of tens to hundreds of megahertz (MHz). Operating in this regime demands using high-frequency techniques, which involve adapting both the experimental set-ups and the mathematical treatment of the acquired data [54]. With low frequency signals the voltage and current present the same value in every point of the circuit at any given moment. However, at high frequencies, if the measuring circuit length is comparable to the wavelength λ of the electromagnetic signal ($\lambda = v/f$, v being the velocity of the light in the transmission medium and f the frequency of the signal), the electric parameters can be different in different parts of the circuit at the same time, and the measured impedance can oscillate between minimum and maximum values depending on the operating frequency. These results are determined mainly by the propagative effects of the signal along the transmission lines, and only very weakly by the actual impedance of the sample under test. Determining the impedance at high frequency requires using specific equipment, such as network analyzers or specialized impedance-measuring systems. The measuring circuit, including the sample holder, should be made of transmission lines (usually, coaxial cables or wave guides). The transmission lines adapted to planar samples are the microstrip and the coplanar lines displayed in **Figure 5**. The microstrip line consists of a narrow strip of flat conductor on a dielectric substrate with the ground conductor in the opposite face. In the coplanar transmission line, the main conductor and the ground are in the top face. In the case of the coplanar line shown in **Figure 5(b)**, both the sample and

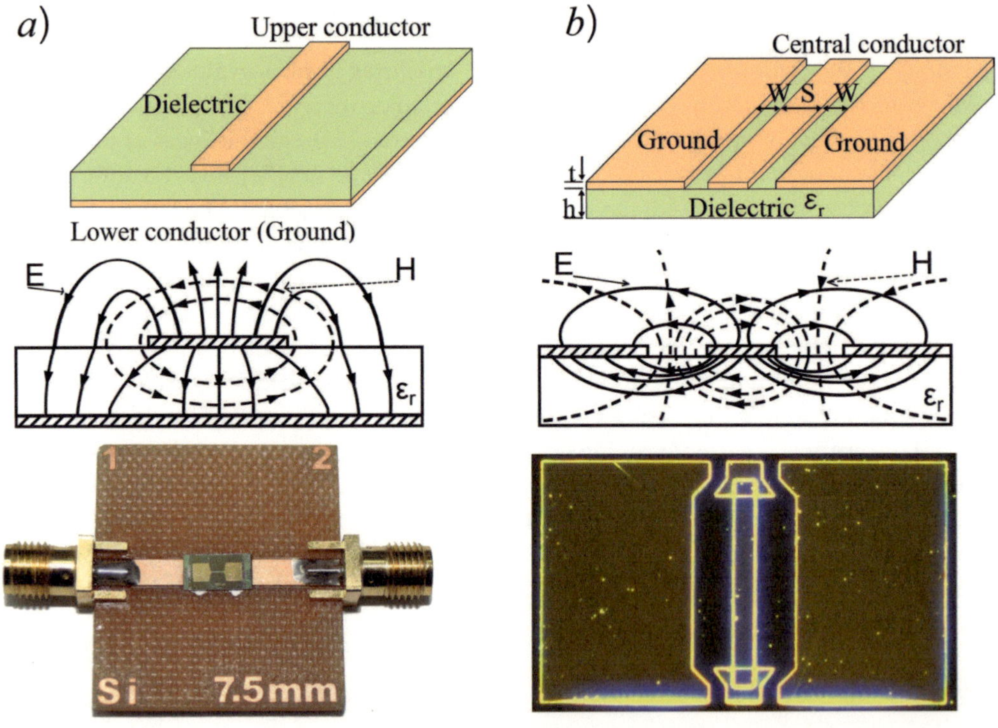

Figure 5. Transmission line-based MI measurement test fixtures. (a) Microstrip line composed by the top conductor and the bottom ground plane, supporting quasi-TEM (transverse electromagnetic mode) waves. The sample is inserted between two portions of line. (b) Coplanar transmission line, with the main conductor and the ground on the topside. In this case, the sample and the conductors of the coplanar line are fabricated by photolithography at the same time.

the line are fabricated by photolithography and the measurements are performed on wafer in a probe station [46]. **Figure 5(a)** displays a usual MI measuring test fixture in which the MI sample is inserted between two sections of a microstrip line, with SMA connectors on both sides to which coaxial cables are attached to connect to the network analyzer.

When measuring with the test fixture displayed in **Figure 5(a)**, standard calibration procedures are employed to eliminate the contribution of the coaxial cables. However, at the measuring frequencies, additional corrections should be applied to eliminate the propagation effects brought about by the portions of the line of the text fixture. The complete details are given in Ref. [54], but **Figure 6** exemplifies how the measured impedance as a function of the frequency is affected by propagation effects and how they can be corrected.

3.2. Impedance as a function of the frequency and the impact of the measuring circuit on MI

MI curves are measured as a function of the magnetic field, which is produced typically by a pair of Helmholtz coils. Using a network (or impedance) analyzer, the real and imaginary parts of the impedance can be measured as a function of the frequency, which can give us a useful insight about the MI effect in thin films. **Figure 7** shows the impedance as a function of the frequency corresponding to a Py/Au/Py sample with transverse anisotropy, measured in a microstrip line (the samples correspond to those of Ref. [55]). The real and imaginary parts of the impedance are plotted at two selected values of the applied magnetic field, $H = 0$, at which the transverse permeability is μ_0, and $H = H^k$, at which the permeability is maximum. First,

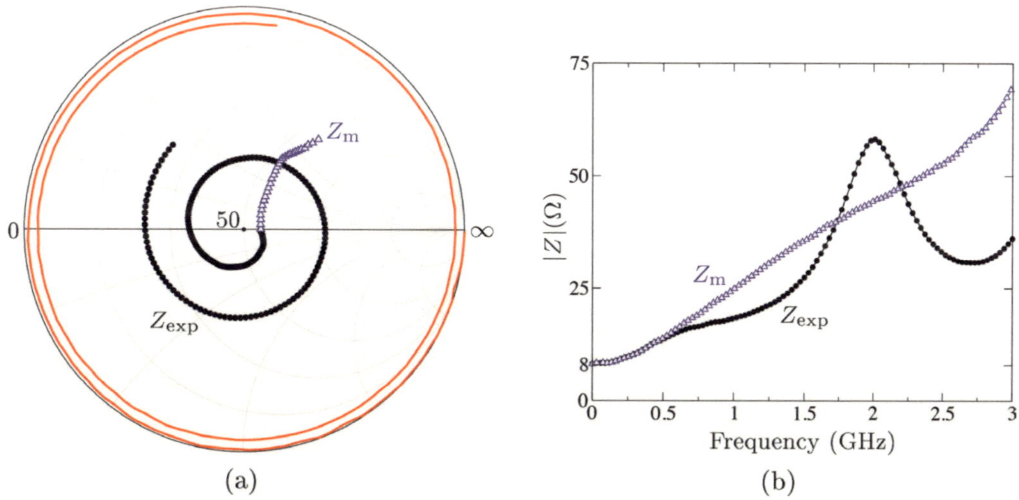

Figure 6. Effect of propagation on impedance measurements and correction. (a) The direct measurement of the impedance of the test fixture with the sample inserted, as a function of the frequency, is plotted on a Smith chart (black circles). Much of the spiral behavior is caused by the portion of the transmission line as observed in the red solid curve, measured with the sample removed from the test fixture. With the suitable correction [54], the impedance of the sample as a function of the frequency can be calculated (blue triangles). (b) The impedance directly measured Z_{exp} (black circles), without corrections, is represented as a function of frequency, showing resonance peaks caused by propagative effects. The corrected impedance Z_m (blue triangles) presents a monotonous behavior.

we show that, although the magneto-inductive effect dominates in sandwiched structures, the skin effect is also important. **Figure 7(a)** shows that the real part at $H = 0$ is constant as a function of the frequency, corresponding to the resistance of the sample (with μ_0, the penetration depth is much large than the thickness of the magnetic layer). In contrast, for $H = H^k$, the real part increases with the frequency due to the skin effect, since the penetration depth is smaller than the thickness of the magnetic layers with large permeability. The relative change of the real part of the impedance keeps growing with the frequency. Second, we observe the extraordinary importance of the external impedance in the performance of MI. The imaginary part of the impedance at $H = 0$ increases linearly from $f = 0$. The slope corresponds to the external inductance of the circuit composed by the sample and the ground plane in the microstrip line [56]. With $H = H^k$, the imaginary part increases hugely for frequencies below 100 MHz, giving a very large impedance change ratio, but afterwards it increases with the same slope as for $H = 0$, because it is dominated by the external inductance that does not depend on the magnitude of the magnetic field. Therefore, the relative change of the imaginary part decreases with frequency. In absolute value, the MI would increase with frequency following the trend

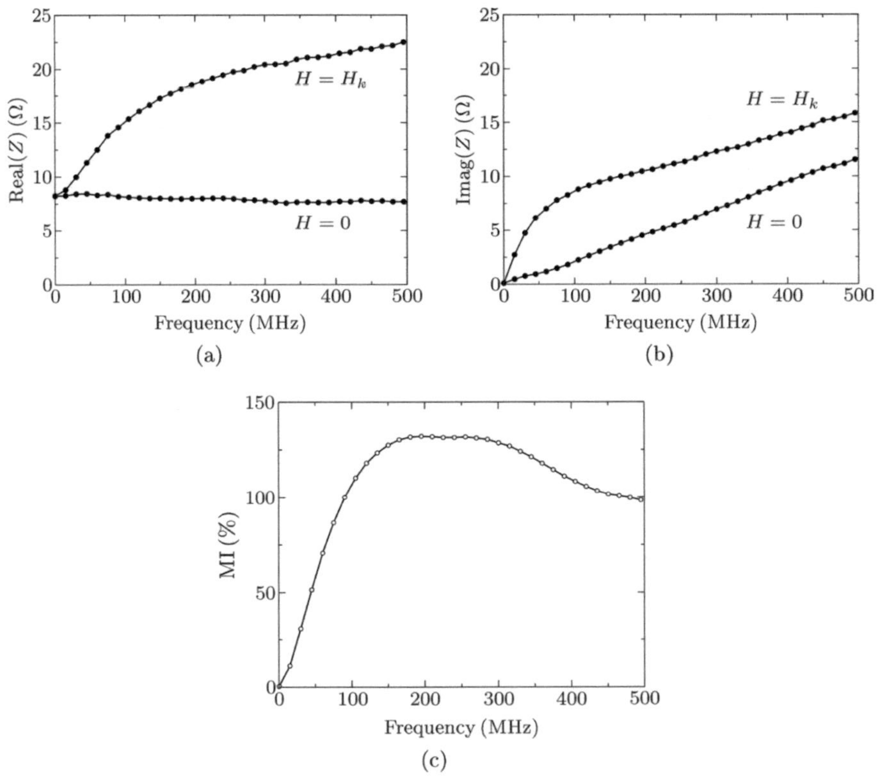

Figure 7. Selected results of the measured impedance as a function of the frequency for a Py/Au/Py multilayer [55]. (a) Real part measured at $H = 0$ and $H = H^k$, where the permeability is large, showing the importance of the skin effect. (b) Imaginary part measured at $H = 0$ and $H = H^k$. The linear increase is caused by the external self-inductance of the measuring circuit. (c) Absolute value of the magneto-impedance. It shows a maximum at approximately the frequency at which the value of imaginary part (ruled by the external inductance) equals the real part: the influence of the impedance added by the measuring circuit limits the maximum MI ratio that can be measured.

of the real part, but it is limited by the decrease of the imaginary part, which is caused by the external inductance. In fact, the maximum of the MI as a function of the frequency takes place at frequencies at which the values of the imaginary and real parts become approximately equal. If the imaginary part is kept low, by reducing the external inductance, the MI can reach larger values, because it would continue growing further with increasing frequency.

It is important to clarify that there is no way of completely avoiding the detrimental effect of the external impedance: for any measurement, or for any application, the sample must be inserted in a circuit that necessarily produces the undesired contribution to the total impedance. Transmission lines are probably the minimum measuring circuit that can be used, but they also produce an important contribution (larger in microstrip than in coplanar lines) that must be taken into account for applications or when analyzing the properties of MI materials and structures. Evidently, the importance of the external contribution in experimental MI measurements explains why the extremely large values of MI predicted by the theory are never reached. As a side effect, we see that the performance rating of materials based purely on their MI ratio is probably meaningless, because the measured values depend heavily on the measuring system. It is possible to apply specialized methods to extract the intrinsic impedance of the sample (a process called de-embedding) for the microstrip [57] and the coplanar cases [46]. This produces much larger MI values and allows, in principle, to compare measured MI values with theoretical predictions.

4. Numerical simulations by finite element method

The MI behavior can be described analytically for simple geometries and, with important assumptions, by expressions similar to Eq. (1) derived from the Maxwell equations. The description can be improved by incorporating models for the magnetization process, the dynamical behavior of the magnetization, and some other effects [58–60]. The pure theoretical analysis allows gaining insight in the involved phenomena, but fails to accurately describe the experimental results. In fact, the theory predicts values for the MI ratio of the order of 10^4% [61, 62] that have never, by far, been observed. This discrepancy is caused first by nonideal conditions on real samples: complex domain structure, finite size, etc. But most importantly, the main cause for this huge difference is that the proposed theories do not reflect the actual measurement conditions. As explained in the previous section, in a real experimental set-up, the measuring circuit imposes a contribution to the measured total impedance that drastically reduces the MI_{max} ratio, even if the circuit is simplified to the minimum expression. A rather complex data reduction process can be used to extract the intrinsic MI response from the sample, eliminating the contribution of the circuit [46, 57], or alternatively, this contribution can be incorporated in the theory [36]. Numerical simulations can overcome some of these problems and offer results closer to the experiment and serve as a guide for device optimization.

Finite element methods (FEMs) are used to solve the differential equations governing many different types of problems in completely arbitrary geometries. For electromagnetic problems, given the material properties and the appropriate boundary conditions, FEM codes can numerically calculate the electrical impedance of the simulated system. If the dependence on the applied magnetic field of the material properties (essentially the magnetic permeability) is specified, FEM can be used to determine the impedance as a function of the magnetic field *Z*(*H*), and the MI ratio according to Eq. (3) [63]. However, to analyze the MI performance of different sample configurations, it is sufficient to use FEM to calculate the impedance for two extreme values of the permeability, the largest μ_{max} (that would correspond to the value of the maximum transverse permeability) and the lowest, which equals to μ_0 when the sample is saturated. This allows to calculate MI_{max} as

$$MI_{max}(\%) = \frac{Z(\mu_{max}) - Z(\mu_0)}{Z(\mu_0)} \times 100 \quad (5)$$

This approach has been used, for example, to examine the performance difference between open- and closed-flux sandwiched structures [20] and to determine the best combination of layer thickness to maximize the MI ratio [53]. The complete three-dimensional (3D) geometry of MI structures can be simulated using commercial FEM software packages as COMSOL [63, 64]. 3D modeling is interesting because it allows, in principle, to account for the effect of shape anisotropy on MI (although no published work has taken advantage of this, yet). If the simulation makes the approximation that the sample is infinite in length, the modeling can be performed in two dimensions (2D), using much less computational resources [65, 66]. The authors routinely use the free 2D software package FEMM [67] to perform FEM simulations. When the simulation cases are numerous or require heavy computing, they use the xfemm implementation of the same code [68] that can be run in a mainframe computer. This FEM program is designed to solve low-frequency electromagnetic problems, in the sense that it implements the Maxwell equations disregarding the displacement current. Therefore, it cannot account for propagation effects and, when solving magnetic problems, it does not consider the dielectric properties of the materials, but it can simulate accurately MI problems in which the skin and magneto-inductive effects are dominant. For a complete simulation of the electromagnetic problem, including propagation and dielectric properties, a *full-wave* program is necessary.

FEM simulations can reproduce quite accurately the experimentally measured MI behavior provided that the simulated problem resembles the experimental conditions. This means that the simulation must include the contribution of the measuring circuit in which the sample is inserted. One of the most popular measuring test fixtures for planar samples, including thin films, is that based on microstrip transmission lines. **Figure 8** schematizes the 2D simulation problem of a thin film structure. Note that, due to the symmetry, only half of the problem must be simulated. The simulation includes, below the sample, the ground conductor of the transmission line, which is made of pure copper, and the dielectric, although it is treated as a dummy material with permeability μ_0, in the same manner as the air filling the rest of the simulation domain. What is important in the simulation is the amount of impedance that the measuring circuit adds to the impedance of the sample. It is mainly produced by its self-inductance, which depends on the geometry of the circuit. The simulation domain must be limited to maintain the problem under reasonable computation demands. Therefore, appropriate boundary conditions

http://dx.doi.org/10.5772/intechopen.70084

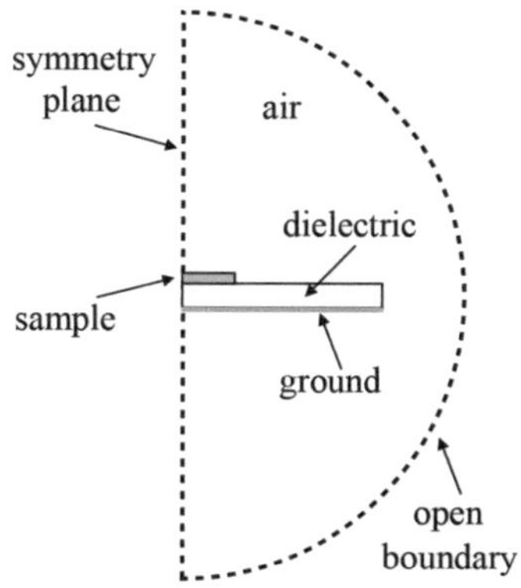

Figure 8. Sketch of the 2D FEM problem for calculating the impedance of a thin film sample (the drawing is not at a real scale). The simulation domain includes the sample and the microstrip line in which it is inserted for real MI measurements. Only half of the problem needs to be simulated thanks to the symmetry. Open boundary conditions are applied to reduce the size of the simulation domain without losing accuracy [67].

must be imposed in the limits of the domain representing the values of the fields over them. This can be problematic in our geometry because the fields are extended, in principle, to the infinite. FEMM offers different possibilities for adequate boundary conditions dealing with that situation, such as the open boundary conditions, represented in **Figure 8**. Other software packages offer similar solutions as the infinite elements used in COMSOL.

FEM calculations are based on discretizing the simulation domain in small portions (triangles in 2D). In principle, the finer the meshing, the more accurate the solution is. However, there is a compromise with computation time and the best strategy is to refine the mesh only where it is important for the solution. Commercial programs include automatic, adaptative, refinement algorithms. In FEMM, this must be done manually. For MI problems, it is especially important to choose a mesh fine enough to accurately determine the fields in the regions where the skin effect is important. As a rule of thumb, we usually chose a mesh 10 times smaller than the penetration depth given by Eq. (1) at each frequency in the appropriate regions.

To calculate the impedance in our FEMM problem, a current is flowed through the sample, perpendicular to the plane of simulation. The current returns in the opposite direction through the ground conductor. Apart from the geometry of the sample, its material properties must be specified. The relevant ones are the conductivity ($\sigma_{Py} = 4.5 \times 10^6$ S/m for Py and $\sigma_{Ti} = 2.38 \times 10^6$ S/m for Ti) and the magnetic permeability. As explained before, the impedance is calculated for two different values of the permeability, μ_{max} and μ_0, assuming the sample is isotropic. This calculation is performed for different frequencies, allowing to obtain MI_{max} versus frequency curves as the ones displayed in **Figure 9**. In this example, the agreement between measurement and simulation is exceptional, up to a frequency of 150 MHz. This degree of coincidence has been secured by performing simulations with different values of μ_{max} until the results matched. FEM simulations therefore constitute a method to determine the maximum value of the transverse permeability of the sample, which is close to $\mu_{max} = 4300\ \mu_0$ in this case. For higher frequencies, there is an increasing discrepancy between simulated and measured values, which can be attributed to the frequency dependence of the permeability.

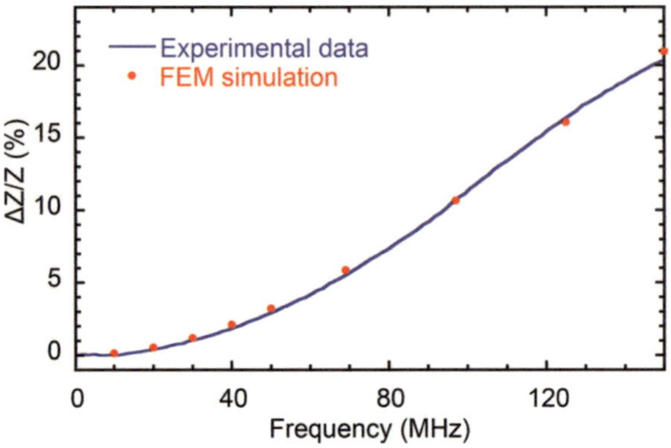

Figure 9. Impedance ratio as a function of the frequency for a thin film multilayer sample [Py (100 nm)/Ti (6 nm)]$_3$/Py (100 nm)]. The agreement between the experimental data and the FEM-simulated values allows determining the value of the transverse permeability of the Py layers.

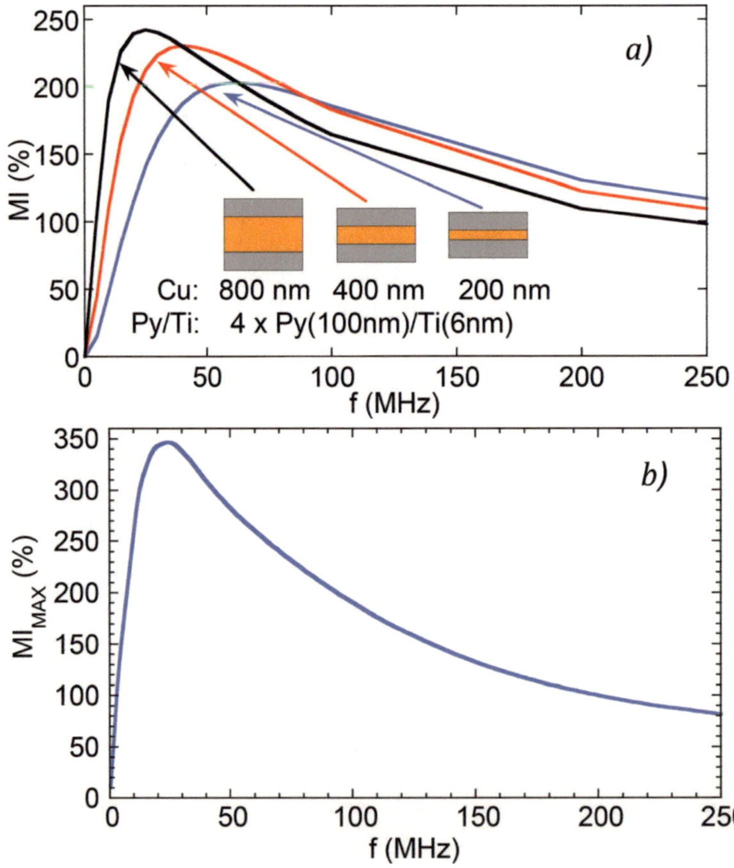

Figure 10. (a) FEM results for sandwiched samples with different thicknesses of the central conductor. The simulations were performed to determine the geometry for maximizing the MI ratio [53]. (b) MI ratio measured in a sample 10 mm long and 0.5 mm wide, fabricated with the optimum structure [Py (100 nm)/Ti (6 nm)]4/Cu (400 nm)/[Ti (6 nm)/Py (100 nm)]$_4$.

In any case, in most situations it is not necessary to exactly match the material parameters to obtain very useful information from FEM simulations. In the work described in Ref. [53], FEM was used to determine the optimum configuration of layers to maximize the MI response. The value of μ_{max} used in the simulations (μ_{max} = 5000 μ_0) was chosen quite arbitrarily but the resulting layer configuration was afterward fabricated, producing record values for both MI ratio and sensitivity for thin film structures with open-flux configuration. If fact, the permeability value used in FEM was underestimated since the measured MI values were larger than those obtained in the simulations. **Figure 10** illustrates these results. It is remarkable that the shape of the measured and simulated curves is, at least qualitatively, identical in the whole frequency range. This could not have been possible without including the measuring circuit (represented by the microstrip elements) in the simulation.

5. Application of thin film MI sensors

5.1. Brief survey of applications

The extremely large sensitivity of the MI effect makes it a good candidate for magnetic sensing applications, comparing favorably with other types of devices as Hall, anisotropic magnetoresistance (AMR), and giant magnetorresistance (GMR) sensors, and competing with fluxgate sensors [69].

MI devices based on amorphous wires have been successfully commercialized by Aichi Corporation as magnetic field sensors for motion, attitude, and acceleration detection [70]. Despite the easier integration of thin film-based structures with microelectronic circuitry, no commercial applications of thin film MI sensors can be found yet. As a pure magnetic field sensor, the application of thin films as an electronic magnetic compass has been evaluated [71].

The MI effect has also been proposed for the detection of magnetic particles, mainly oriented toward possible biomedical applications. Those were exhaustively reviewed recently [72]. Among the different materials used, besides wires and ribbons, thin film structures have been demonstrated to be able to detect microparticles in a flowing liquid [73] and nanoparticles embedded in gels (ferrogel) [74]. In these works, the presence of magnetic particles modifies the magnetic permeability on the medium surrounding the MI element, producing the measured changes in the impedance. Note that there is no direct measurement of the fringe field created by the particles, since no external magnetic field is applied to magnetize them. The field used to bias the sensing element (H_B in **Figure 1**) is not enough for that end. In fact, the electrical impedance of nonmagnetic conductors also varies in the presence of magnetic particles and this effect has been proposed as an alternative method for biosensing [75]. More sensitivity is expected if the fringe field of the particles is measured directly using the MI effect as a magnetic field detector. There is no inconvenient, in principle, in using thin film sensors substituting magnetic wires in the set-up described in Ref. [76].

MI has also been proposed as sensing mechanism for nondestructive testing (NDT), using amorphous wires [77] and thin film structures [78, 79], and even for pressure sensors using thin films deposited on a flexible substrate [40]. In fact, the use of flexible substrates for thin film MI implementations is gaining relevance with the onset of applications of flexible electronics [39–42]. Novel deformation sensors are being proposed based on stress-impedance effect, which is very close to the MI effect, but based on the variations of the magnetic permeability experienced by the sensing material upon the application of mechanical strains [80, 81].

5.2. Circuit conditioning: impedance matching

The ultimate goal of the application of the MI effect is to develop a device that can compete in some aspect (sensitivity, miniaturization, prize, etc.) with the family of magnetic sensors that are widely used nowadays. Obviously, the circuit architecture in an industrially produced MI sensor would differ from the set-up implemented in the laboratory experiments. One of the aspects that would be benefited from a different circuit approach in a commercial device is the fact that the load impedance of the sensing element should be matched to the characteristic impedance of the transmission line, typically 50 Ω. The main goal of this procedure is to minimize the power consumption of the electronics, but in the case of a MI sensor it produces the additional benefit of an increased sensitivity. Matching the circuit load to 50 Ω is achieved by introducing impedance-matching networks in the circuit, typically consisting of L and C elements whose values are calculated to drive the impedance of the sample at the operation point (in general, complex) to 50 Ω (purely resistive) [82]. Thus, the imaginary part of the impedance of the sample gets cancelled with the addition of the LC network. As explained in Section 3.2, the imaginary part of the impedance is much less sensitive than the real part, due to the contribution of the external inductance. Therefore, by cancelling it and operating effectively with a purely resistive device, the MI ratio and sensitivity of the sample get noticeably improved. An example of this procedure applied to MI thin films, together with detailed calculations and results, can be found in Ref. [83].

5.3. Magnetic noise in thin film MI sensors

The parameter that really allows a faithful performance comparison between different magnetic sensors is their level of magnetic noise, which determines the threshold for the minimum signal level that can be reliably detected. Noise levels below 1 pT/√Hz can be obtained in MI sensors based on amorphous wires and ribbons [84, 85], but very little work has been done on determining the noise level in thin film-based structures. In a real application, the MI element is inserted in a conditioning circuit that, in its simplest configuration, is composed of a RF oscillator (that produces the ac current exciting the MI sensitive element), a detector, and an amplifier [86]. Each of these elements contributes to the total noise of the device. In an attempt to determine the noise of a thin film structure (a sandwiched multilayer 2 mm long and 90 μm wide with a sensitivity of 55%/Oe), the magnetic noise has been measured using different detectors: a commercial AD8302 Gain and Phase detector, a commercial power detector ZX47-50+, and a homemade peak detector [87]. The impedance of the sample is matched to 50 Ω using a specific matching network, to increase its sensitivity, as explained in the previous section. The measured electrical noise e_n is transformed in magnetic noise b_n expressed in T/√Hz through the experimentally determined transfer function of the sensor, that is, the relation between the input (magnetic field) and the output (voltage signal) of the device. The sensor using the peak detector displayed the lowest noise level of 3 nT/√Hz at 1 Hz and 122 pT/√Hz at 2 kHz, as shown in **Figure 11**.

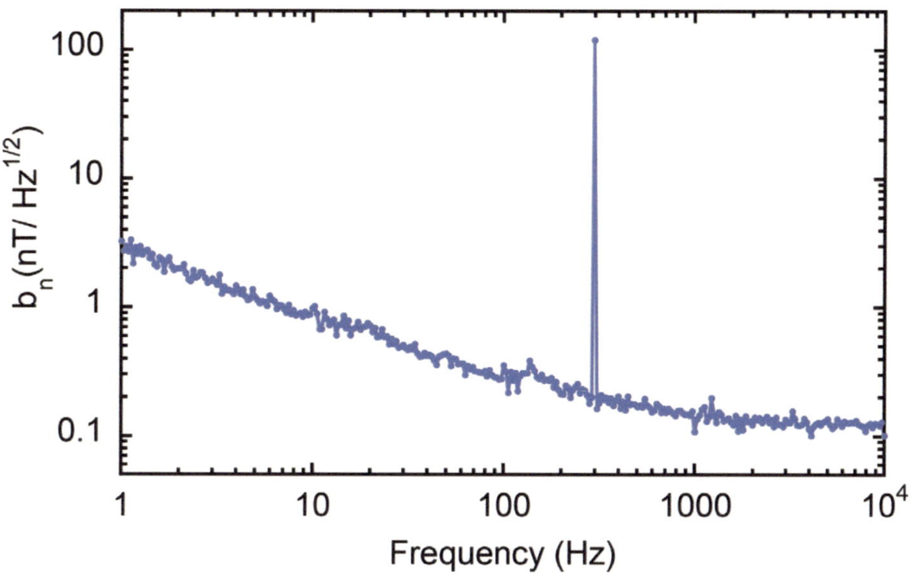

Figure 11. Equivalent magnetic noise measured in a thin film sandwiched multilayer sample, 2 mm long and 90 μm wide with structure [Py (170 nm)/Ti (6 nm)]$_3$/Cu (250 nm)/[Ti (6 nm)(Py (170 nm)]$_3$.

These noise levels are significantly larger than the ones reported in wires and ribbons [84, 85], but are promising for applications. Besides, the noise level can be reduced easily by a factor of 4 by using MI thin films with enhanced sensitivities (up to 300%/Oe), as the ones described recently [53]. Additionally, lower total noise values are expected by minimizing the RF oscillator contribution to noise [88].

6. Conclusions

Thin film structures displaying the magneto-impedance effect are already mature for sensing applications. Once their performance has reached levels comparable to those of amorphous wires, their fabrication compatibility with the microelectronic production processes makes them attractive for integrated devices. We have briefly reviewed the main concepts involved in the development of MI materials for future sensors. Many more details can be found in the proposed references, especially in the thesis reports by David de Cos [56] and Eduardo Fernández [88], on which much of the presented work is based.

Acknowledgements

Parts of the work summarized here have been performed thanks to the financial support from the Basque Government under grants KK-2016/00097 (Actimat Project) and KK-2016/00030 (Micro4Fab Project).

Author details

Alfredo García-Arribas[1,2]*, Eduardo Fernández[2,3] and David de Cos[4]

*Address all correspondence to: alfredo.garcia@ehu.es

1 Department of Electricity and Electronics, University of the Basque Country UPV/EHU, Leioa, Spain

2 Basque Centre for Materials, Applications and Nanostructures BCMaterials, Bizkaia Science and Technology Park, Derio, Spain

3 Department of Materials Science and Engineering, Massachusetts Institute of Technology, Cambridge, USA

4 Department of Applied Physics II, Faculty of Pharmacy, University of the Basque Country UPV/EHU, Vitoria, Spain

References

[1] Beach RS, Berkowitz AE. Giant magnetic field dependent impedance of amorphous FeCoSiB wire. Applied Physics Letters. 1994;**64:**3652

[2] Panina LV, Mohri K. Magneto-impedance effect in amorphous wires. Journal of Applied Physics. 1994;**65**:1189

[3] Lord Rayleigh. On the behaviour of iron and steel under the operation of feeble magnetic forces. Philosophical Magazine. 1887;**23**:225

[4] Kittel C. Theory of the dispersion of magnetic permeability in ferromagnetic materials at microwave. Physical Review. 1946;**70**:281

[5] Sommerfeld A. Electrodynamics. New York: Academic; 1952. p. 166

[6] Landau LD, Lifshitz EM. Electrodynamics of Continuous Media. Oxford: Pergamon; 1975. p. 211

[7] Chen D-X, Muñoz JL. AC impedance and circular permeability of slab and cylinder. IEEE Transactions on Magnetics. 1999;**35**:1906

[8] Cullity BD. Introduction to Magnetic Materials. 2nd ed. Menlo Park, California: Wiley-IEEE Press; 2009. p. 226

[9] Vázquez M. Soft magnetic wires. Physica B. 2001;**299**:302

[10] Vázquez M. Giant magneto-impedance in soft magnetic "wires". Journal of Magnetism and Magnetic Materials. 2001;**226-230**:693

[11] Vázquez M, Hernando A. A soft magnetic wire for sensor applications. Journal of Physics D: Applied Physics. 1996;**29**:939

[12] Machado FLA, Rezende SM. A theoretical model for the giant magnetoimpedance in ribbons of amorphous soft-ferromagnetic alloys. Journal of Applied Physics. 1996;**79**:6558

[13] Hernando B, Sánchez ML, Prida VM, Tejedor M, Vázquez M. Magnetoimpedance effect in amorphous and nanocrystalline ribbons. Journal of Applied Physics. 2001;**90**:4783

[14] Hika K, Panina LV, Mohri K. Magneto-impedance in sandwich film for magnetic sensor heads. IEEE Transactions on Magnetics. 1996;**32**:4594

[15] Morikawa T, Nishibe Y, Yamadera H, Nonomura Y, Takeuchi M, Taga Y. Giant magneto-impedance effect in layered thin films. IEEE Transactions on Magnetics. 1997;**33**:4367

[16] Antonov A, Gadetsky S, Granovsky A, D'yatckpv A, Sedova M, Perov N, Usov N, Furmanova T, Lagar'kov A. High-frequency giant magneto-impedance in multilayered magnetic films. Physica A. 1997;**241**:414

[17] Kurlyandskaya GV, Yakabchuck H, Kisker E, Bebenin NG, García-Miquel H, Vázquez M, Vas'kovskiy VO. Very large magnetoimpedance effect in FeCoNi ferromagnetic tubes with high order magnetic anisotropy. Journal of Applied Physics. 2001;**90**:6280

[18] Gromov A, Korenivski V. Electromagnetic analysis of layered magnetic/conductor structures. Journal of Physics D: Applied Physics. 2000;**33**:773

[19] Panina LV, Makhnovskiy DP, Maps DJ, Zarechnyuk DS. Two-dimensional analysis of magnetoimpedance in magnetic/metallic multilayers. Journal of Applied Physics. 2001; **89**:7221

[20] Fernández E, García-Arribas A, Volchkov SO, Kurlyandskaya GV, Barandiaran JM. Differences in the magneto-impedance of FeNi/Cu/FeNi multilayers with open and closed magnetic path. IEEE Transactions on Magnetics. 2010;**46**:658

[21] Morikawa T, Nishibe Y, Yamadera H, Nonomura Y, Takeuchi M. Sakata J, Taga Y. Enhancement of giant magneto-impedance in layered film by insulator separation. IEEE Transactions on Magnetics. 1996;**32**:4965

[22] Yu J, Zhou Y, Cai B, Xu D. Giant magneto-impedance effect in amorphous magnetostrictive FeSiB thin films. Journal of Magnetism and Magnetic Materials. 2000;**213**:32

[23] Nishibe Y, Yamadera H, Ohta N, Tsukada K, Ohmura Y. Magneto-impedance effect of a layered CoNbZr amorphous film formed on a polyimide substrate. IEEE Transactions on Magnetics. 2003;**39**:571

[24] Correa MA, Bohn F, Viegas ADC, Carara MA, Schelp LF, Sommer RL. Giant magneto-impedance in FM/SiO_2/Cu/SiO_2/FM films at GHz frequencies. Journal of Magnetism and Magnetic Materials. 2008;**320**:e25

[25] Tumanski S. Thin Film Magnetoresistive Sensors. London: IoP publishing; 2001

[26] Sugita Y, Fujiwara H, Sato T. Critical thickness and perpendicular anisotropy of evaporated films with stripe domains. Applied Physics Letters. 1967;**10**:229

[27] Coïsson M, Vinal F, Tiberto P, Celegato F. Magnetic properties of FeSiB thin films displaying stripe domains. Journal of Magnetism and Magnetic Materials. 2009;**321**:806

[28] Svalov AV, Aseguinolaza IR, García-Arribas A, Orue I, Barandiaran JM, Alonso J, Fernández-Gubieda ML, Kurlyandskaya GV. Structure and magnetic properties of thin film Permalloy films near the "transcritical" state. IEEE Transactions on Magnetics. 2010;**46**:333

[29] Corrêa MA, Viegas ADC, da Silva RB, de Andrade AMH, Sommer RL. Magnetoimpedance of single and multilayered FeCuNbSiB films in frequencies up to 1.8 GHz. Journal of Applied Physics. 2007;**101**:043905

[30] Kurlyandskaya GV, Svalov AV, Fernández E, García-Arribas A, Barandiaran JM. Journal of Applied Physics. 2010;**107**:09C502

[31] Svalov AV, Fernández E, García-Arribas A, Alonso J, Fdez-Gubieda ML, Kurlyandskaya GV. Applied Physics Letters. 2012;**100**:162410

[32] Kurlyandskaya GV, García-Arribas A, Fernández E, Svalov AV. Nanostructured magnetoimpedance multilayers. IEEE Transactions on Magnetics. 2012;**48**:1375

[33] Ohmori K, Tan K, Itoi K, Nagasu K, Uemichi Y, Aizawa T, Yamauchi R. Thin film magneto-impedance sesnor integrated into $L1_0$ FePt thin film bias magnet. IEEE Transactions on Magnetics. 2008;**44**:3977

[34] García C, Florez JM, Vargas P, Ross CA. Asymmetrical giant magnetoimpedance in exchange-biased NiFe. Applied Physics Letters. 2010;**96**:232501

[35] Svalov A, Savin P, Lepalovskij V, Larrañaga A, Vas'kovskiy V, García-Arribas A, Kurlyandskaya G. Tailoring the exchange bias in FeNi/FeMn bilayers by heat treatment and FeMn surface oxidation. IEEE Transactions on Magnetics. 2014;**50**:2003904

[36] Corrêa MA, Bohn F, da Silva RB, Sommer RL. Magnetoimpedance effect at the high frequency range for the thin film geometry: Numerical calculation and experiment. Journal of Applied Physics. 2014;**116**:243904

[37] Wang T, Lei C, Lei J, Yang Z, Zhou Y. Preparation of meander thin-film microsensor and investigation of structural parameters on the giant magnetoimpedance effect. Applied Physics A. 2012;**109**:205

[38] Takayama A, Umehara T, Yuguchi A, Kato H, Mohri K, Uchiyama T. Integrated thin film magneto-impedance sensor head using plating process. IEEE Transactions on Magnetics. 1999;**35**:3643

[39] Kurlyandskaya GV, Fernández E, Svalov A, Burgoa Beitia A, García-Arribas A, Larrañaga A. Flexible thin film magnetoimpedance sensors. Journal of Magnetism and Magnetic Materials. 2016;**415**:91

[40] Fernández E, Kurlyandskaya GV, García-Arribas A, Svalov AV. Nanostructured giant magneto-impedance multilayers deposited onto flexible substrates for low pressure sensing. Nanoscale Research Letters. 2012;**7**:230

[41] Agra K, Mori TJA, Dorneles LS, Escobar VM, Silva UC, Chesman C, Bohn F, Corrêa MA. Dynamic magnetic behavior in non-magnetostrictive multilayered films grown on glass and flexible substrates. Journal of Magnetism and Magnetic Materials. 2014;**355**:136

[42] Li B, Kavaldzhiev MN, Kosel J. Flexible magnetoimpedance sensor. Journal of Magnetism and Magnetic Materials. 2015;**378**:499

[43] García-Arribas A, Fernández E, Barrainkua A, Svalov AV, Kurlyandskaya GV, Barandiaran JM. Comparison of micro-fabrication routes for magneto-impedance elements: Lift-off and wet-etching. IEEE Transactions on Magnetics. 2012;**48**:1601

[44] Giouroudi I, Hauser H, Musiejovsky L, Steurer J. Development of amorphous thin film meander trilayers and investigation of GMI effect. Proceedings of IEEE Sensors 2004. 2004;**2**:1024

[45] Kikuchi H, Oe S, Uetake H, Yabukami S, Nakai T, Hashi S, Ishiyama K. Enhancement of sensitivity on miniaturized thin-film magnetoimpedance with ellipsoidal element. Physics Procedia. 2015;**75**:1271

[46] Fernández E, López A, García-Arribas A, Svalov AV, Kurlyandskaya GV, Barrainkua A. High-frequency magnetoimpedance response of thin-film microstructures using coplanar waveguides. IEEE Transactions on Magnetics. 2015;**51**:1

[47] Fernández E, Svalov AV, Kurlyandskaya GV, García-Arribas A. GMI in nanostructured FeNi/Ti multilayers with different thicknesses of the magnetic layers. IEEE Transactions on Magnetics. 2013;**49**:18

[48] Volchkov SO, Fernández E, García-Arribas A, Barandiaran JM, Lepalovskij VN, Kurlyandskaya GV. Magnetic properties and giant magnetoimpedance of FeNi-based nanostructured multilayers with variable thickness of the central Cu lead. IEEE Transactions on Magnetics. 2011;**47**:3328

[49] García-Arribas A, Fernández E, Svalov AV, Kurlyandskaya GV, Barrainkua A, Navas D, Barandiaran JM. Tailoring the magnetic anisotropy of thin film permalloy microstrips by combined shape and induced anisotopropies. European Physical Journal B. 2013;**86**:136

[50] Kikuchi H, Kamata S, Nakai T, Hashi S, Ishiyama K. Influence of demagnetizing field on thin-film GMI magnetic sensor elements with uniaxial magnetic anisotropy. Sensors and Actuators B. 2015;**230**:142

[51] Makhnovskiy DP, Panina LV. Size effect on magneto-impedance in layered films. Sensors and Actuators A. 2000;**81**:91

[52] García-Arribas A, Fernández E, Orue I, Barandiaran JM. Determination of the distribution of transverse magnetic anisotropy in thin films from the second harmonic of Kerr signal. Applied Physics Letters. 2013;**103**:142411

[53] García-Arribas A, Fernández E, Svalov A, Kurlyandskaya GV, Barandiaran JM. Thin-film magneto-impedance structures with very large sensitivity. Journal of Magnetism and Magnetic Materials. 2016;**400**:321

[54] de Cos D, García-Arribas A, Barandiaran JM. Analysis of magnetoimpednace measurements at high frequency using a microstrip transmission line. Sensors and Actuators A. 2004;**115**:368

[55] de Cos D, Fry N, Orue I, Panina LV, García-Arribas A, Barandiaran JM. Very large magnetoimpedance (MI) in FeNi/Au multilayer film systems. Sensors and Actuators A. 2006;**129**:256

[56] de Cos D. Magento-impedancia de muestras planas en alta frecuencia (in spanish) [thesis]. Leioa, Spain: Universidad del País Vasco, UPV/EHU; 2006

[57] García-Arribas A, de Cos D, Barandiaran JM. Determination of the intrinsic high-frequency magnetoimpedance spectra of multilayer systems. Journal of Applied Physics. 2006;**99**:08C507

[58] Makhnovskiy DP, Panina LV, Mapps DJ. Field-dependent surface impedance tensor in amorphous wires with two types of magnetic anisotropy: Helical and circumferential. Physical Review B. 2001;**63**:144424

[59] Kraus L. Theory of giant magneto-impedance in the planar conductor with uniaxial magnetic anisotropy. Journal of Magnetism and Magnetic Materials. 1999;**195**:764

[60] Barandiaran JM, García-Arribas A, Muñoz JL, Kurlyandskaya GV, Valenzuela R. Domain wall permeability limit for the giant magnetoimpedance effect. Journal of Applied Physics. 2002;**91**:7451

[61] Kraus L. The theoretical limits of giant magneto-impedance. Journal of Magnetism and Magnetic Materials. 1999;**196**:354

[62] Barandiarán JM, García-Arribas A, Muñoz JL, Kurlyandskaya GV. Influence of magnetization processes and device geometry on the GMI effect. IEEE Transactions on Magnetics. 2002;**38**:3051

[63] Nejad SN, Mansour R. A 3D finite-element analysis of giant magnetoimpedance thin-film magnetic sensors. IEEE Transactions on Magnetics. 2016;**52**:4000208

[64] Li B, Kosel J. Three dimensional simulation of giant magneto-impedance effect in thin film structures. Journal of Applied Physics. 2011;**109**:07E519

[65] Sinnecker JP, Pirota KR, Knobel M, Kraus L. AC magnetic transport on heterogeneous ferromagnetic wires and tubes. Journal of Magnetism and Magnetic Materials. 2002;**249**:16

[66] García-Arribas A, Barandiarán JM, de Cos D. Finite element method calculations of GMI in thin films and sandwiched structures: Size and edge effects. Journal of Magnetism and Magnetic Materials. 2008;**320**:e4

[67] Meeker DC. Finite Element Method Magnetics, Version 4.2 [Internet]. 12 January 2016 Build. Available from: http://www.femm.info

[68] Crozier R, Mueller M. A New MATLAB and Octave interface to a popular magnetics finite element code. Proceedings of the 22nd International Conference on Electric Machines (ICEM 2016); 4-7 Sept. 2016; Lausanne, Switzerland. IEEE; 2016

[69] Ripka P, Janosek M. Advances in magnetic field sensors. IEEE Sensors Journal. 2010;**10**:1108

[70] Mohri K, Uchiyama T, Panina L V, Yamamoto M, and Bushida K, Recent Advances of Amorphous Wire CMOS IC Magneto-Impedance Sensors: Innovative High-Performance Micromagnetic Sensor Chip. Journal of Sensors. 2015;**2015**:718069

[71] García-Arribas A, Fernández E, Orue I, de Cos D, Barandiaran JM, Michelena JM, Martínez F. Evaluation of a thin film giant magneto-impedance electronic compass. Sensor Letters. 2013;**11**:36

[72] Wang T, Zhou Y, Lei C, Luo J, Xie S, Pu H. Magnetic impedance biosensor: A review. Biosensors and Bioelectronics. 2017;**90**:418

[73] García-Arribas A, Martínez F, Fernández E, Ozaeta I, Kurlyandskaya GV, Svalov AV, Berganzo J, Barandiaran JM. GMI detection of magnetic-particle concentration in continuous flow. Sensors and Actuators A. 2011;**172**:103

[74] Kurlyandskaya GV, Fernández E, Safronov AP, Svalov AV, Beketov I, Burgoa Beitia A, García-Arribas A, Blyakhman FA. Giant magnetoimpedance biosensor for ferrogel detection: Model system to evaluate properties of natural tissue. Applied Physics Letters. 2015;**106**:193702

[75] Lago-Cachón D, Rivas M, Martínez-García JC, García JA. Cu impedance-based detection of superparamagnetic nanoparticles. Nanotechnology. 2013;**24**:245501

[76] Fodil K, Denoual M, Dolabdjian C, Treizebre A, Senez V. In-flow detection of ultra-small magnetic particles by an integrated giant magnetic impedance sensor. Applied Physics Letters. 2016;**108**:173701

[77] Vacher F, Alves F, Gilles-Pascaud C. Eddy current nondestructiuve testing with giant magneto-impedance sensor. NDT&E International. 2007;**40**:439

[78] Barandiaran JM, Kurlyandskaya GV, de Cos D, García-Arribas A, Vas'kovskiy VO. Multilayer magnetoimpedance sensor for nondestructive testing. Sensor Letters. 2009;**7**:374

[79] Moulin J, Dufourg-Gergam E, Oubensaid E-H, Alves F, Kaviraj B, Gupta A, Reddy VR, Barrué R. Realisation of micro-patterned multi-element magneto-impedance sensors for non destructive testing. Sensor Letters. 2009;**7**:272

[80] Agra K, Bohn F, Mori TJA, Callegari GL, Dorneles LS, Correa MA. Handling magnetic anisotropy and magnetoimpedance effect in flexible multilayers under external stress. Journal of Magnetism and Magnetic Materials. 2016;**420**:81

[81] García-Arribas A, Combarro L, Goiriena-Goikoetxea M, Kurlyandskaya GV, Svalov AV, Fernández E, Orue I, Feuchtwanger J. Thin-film magnetoimpedance structures onto flexible substrates as deformation sensors. IEEE Transactions on Magnetics. 2017;**53**:2000605

[82] Bowick C. RF Circuit Design. 1st ed. Indianapolis: Howards Sms & Co; 1982

[83] de Cos D, Sandacci SI, García-Arribas A, Barandiaran JM. Impedance matching networks for power transfer and sensitivity enhancement in GMI sensors. IEEE Transactions on Magnetics. 2005;**41**:3655

[84] Dufay B, Saez S, Dolabdjian CP, Yelon A, Menard D. Characterization of an optimized off-diagonal GMI-based magnetometer. IEEE Sensors Journal. 2013;**13**:379

[85] Malatek M, Dufay B, Saez S, Dolabdjian C. Improvement of the off-diagonal magneto-impedance sensor white noise. Sensors and Actuators A. 2013;**204**:20

[86] Ding L, Saez S, Dolabdjian C, Melo LGC, Yelon A, Ménard D. Equivalent magnetic noise limit of low-cost GMI magnetometer. IEEE Sensors Journal. 2009;**9**:159

[87] Fernández E, García-Arribas A, Barandiaran JM, Svalov AV, Kurlyandskaya GV, Dolabdjian CP. Equivalent magnetic noise of micro-patterned multilayer thin films based GMI microsensor. IEEE Sensors Journal. 2015;**15**:6707

[88] Fernández E. Thin-film magnetoimpedance micro-structures for sensing applications [thesis]. Leioa, Spain: Universidad del País Vasco, UPV/EHU; 2016

Section 2

Magneto-Resistance (MR) Sensors

Chapter 4

Engineering Magnetoresistance in Mn_xGe_{1-x} System for Magnetic Sensor Application

Tianxiao Nie, Weisheng Zhao and Kang L. Wang

Additional information is available at the end of the chapter

http://dx.doi.org/10.5772/intechopen.70206

Abstract

In 2007, a Nobel Prize is awarded to Dr. Albert Fert and Peter Grünberg for their contribution in giant magnetoresistance (GMR) effect. The magnetic head based on GMR effect has significantly increased the storage density in the hard disk drive (HDD) and brought the coming of the digital age. Besides, the rapid development of GMR sensor has opened a wide and promising range of applications, including the aspects in automobile, traffic monitor, biomedicine, and space, etc. As continuously extending the market, it needs GMR sensor with much lower cost, smaller size, higher sensitivity, and compatibility with the CMOS technology. In light of that, we give a review about the recent progress of the MR effect in Mn_xGe_{1-x} system, which refers to the material growth and magnetic and MR property. Through engineering the Mn_xGe_{1-x} structure, it could realize the transition from negative to positive MR, geometric-enhanced giant MR, and electric-field controlled MR. The fact of well-designed MR effect and high compatibility with Si technology brings a high potential and advantage for fabricating Mn_xGe_{1-x}-based MR sensors, which could be widely used in magnetic head and biomedical sensors, among others, with the superiority of much lower manufacturing cost, lower power dissipation, higher integration density, and higher sensitivity.

Keywords: magnetoresistance, geometric-enhanced MR, electric-field controlled MR, Mn_xGe_{1-x} system, MR sensor

1. Introduction

Magnetic sensors have been widely used in analyzing and controlling thousands of functions by human beings for many decades [1–3]. A digital world has arrived and been driven by the tremendous increase of storage density in the hard disk drive (HDD) using the state-of-the-art magnetic sensor [4]. The convenient transportation has been promised in a safe manner because of the high reliability of the noncontact switching and monitor with magnetic sensor [5–7]. The portable medical system relies on the magnetic sensor for nondestructive diagnostic applications [8–10]. Factories have high productivity due to the high stability and precision as well as low cost of magnetic sensors [3, 11].

There are numerous types of magnetic sensors [6], such as search-coil sensors, fluxgate sensors, magneto-optical sensors, optically pumped magnetometer, magnetoresistance (MR) sensors, Hall effect sensors, and superconducting quantum interference device (SQUID) magnetometer, among others. Most of them are based on the direct magnetic and electric response. A magnetic sensor can directly convert the magnetic signal into an inductive voltage signal [12] or resistance variation [5], and its sensitivity determines its operating regime and potential applications. For example, SQUID magnetometer with a high sensitivity of 10^{-10}–10^{-4} gauss has been used for measuring magnetic field gradients or differences due to permanent dipole magnets in major applications of brain function mapping. The fluxgate and MR sensor can provide the medium field sensitivity of 10^{-6}–10^{2} gauss, which has been used for the magnetic compass and magnetic anomaly detection. Hall effect sensors with a low sensitivity of 1–10^{6} gauss have been exploited for applications in noncontact switching and current meters. In consideration of the operating regime, the MR sensor is one of the most commonly used sensors in everyday life [4, 5, 11, 13, 14]. Additionally, it demonstrates some specific advantages [4, 5] comparing to others, including high compatibility with the CMOS process technology, high scalability, low power dissipation, and low manufacture cost.

MR sensors using the resistance as the detectable signal involve the contributions from different mechanisms, such as anisotropic magnetoresistance (AMR) [5], giant magnetoresistance (GMR) [5, 15, 16], and tunnel magnetoresistance (TMR) [17, 18]. AMR sensor could exhibit large field sensitivity and more significantly could detect the field direction, which has been widely used as the magnetic head in HDD and automotive sensors to determine many quantities, like throttle valve position, chassis height, and pedal position [5]. However, since the discovery of GMR effect in 1988 [15], it has gradually replaced the role of the AMR effect for the sensor application, which provided more advantages, such as larger output signal, miniaturization opportunity, and the possibility to make a 360° angle sensor. As a derivative of GMR effect, TMR effect could have much higher sensitivity and integration density [18]. A successful paradigm for sensor evolution is the magnetic read head. A dramatic shift to storing data from the analog to current digital world started in the 2000s and was dominantly driven by the emergence of spintronics exemplified by the introduction of the AMR read head in 1991 by IBM [19], the GMR read head in 1997 [20, 21], and the TMR read head in 2006 [21]. Prior to the introduction of the MR read head, the storage densities of HDD had increased at a growth rate of 25% per year. With the introductions of the AMR head in 1991, the growth rate of storage density increased to 60% per year, while the introduction of GMR read head made the growth rate further to 100% per year. Due to such dramatic impact, the 2007 Nobel Prize in Physics was awarded to Albert Fert and Peter Grünberg for the discovery of GMR effect. The recent areal density growth has slowed resulting mainly from the thermal instabilities, known as the superparamagnetic effect [22, 23]. While the continual effort attempts to address the fundamental limitation of GMR sensor for much higher sensitivity and thermal instability, searching for other high-efficient MR sensors has never been stopped.

As a potential candidate, the geometric-enhanced MR could demonstrate a large MR value as a function of magnetic field, and the underlying mechanism arises from the dependence of the current path on the sample shape and electrode configuration [24]. GMR effect can give rise to a negative MR, whereas the geometric-enhanced MR usually is positive. The current deflected from the electric-field direction by Lorentz force, and the inhomogeneity in current path determined the MR sensitivity. Previously, Thio et al. reported a large MR up to 28% at 500 gauss in the inhomogeneous $Hg_{1-x}Cd_xTe$ thin film associated with the composition fluctuation [24, 25]. Furthermore, Solin et al. configured an inhomogeneous structure by embedding concentric gold in nonmagnetic indium antimonide matrix and showed a MR value as high as 100% at 500 gauss [24]. All of these results point to a new way for designing new type of MR sensors, and the related research is still in progress.

As the demand for the sensors with higher integration density and lower manufacturing cost is continuously growing, it calls for not only innovative sensor configuration but also novel material candidate with much higher sensitivity to magnetic field, higher compatibility with CMOS technology, and lower power consumption. Mn_xGe_{1-x} [26] as the silicon-compatible material appears to be an appealing candidate for application in MR sensors through the combined use of GMR [27] and geometric-enhanced MR [28]. In light of that, we give a review of the current research progress in Mn_xGe_{1-x} thin film and nanostructures, as well as their possible application in MR sensors. The fundamental aspects of the MR mechanisms are listed in the first section. The following section gives a systematic and comprehensive review about the Mn_xGe_{1-x} material growth, structure characterization, and MR measurement. MR phenomena based on different mechanisms including GMR, geometric-enhanced MR, and even electric-field controlled MR have been well discussed, and the correlation between the structures and MR properties is established. The final section gives an outlook of the potential application of the Mn_xGe_{1-x}-based MR sensors and estimates their implicit impact. Exploiting Mn_xGe_{1-x}-based MR sensors may set a new stage for the next-generation sensors with improved sensitivity, higher scalability, and higher compatibility with current CMOS technology.

2. The fundamental principles in MR sensors

2.1. The definition of MR effect

The definition of the MR comes from the resistance variation of a material as a function of magnetic field, which can be described by the following equation:

$$MR = \frac{R(B) - R(0)}{R(0)} = f(B) \tag{1}$$

where the R(B) and R(0) are the resistance at the magnetic field of B and zero, respectively. The MR value follows different functions with the magnetic field based on the mechanism differences. In a traditional semiconductor, the MR abides by the orbital MR effect with the origin from Lorentz force. The deflection of the current due to the magnetic field produces an increase of the current path length and thus an increase of resistance. The relation between them can be described as:

$$MR = \frac{\rho_B}{\rho_0}(1 + C_1 (\mu B)^2) \qquad (2)$$

where ρ_B/ρ_0 is the specific relative resistance, C_1 is a geometrical parameter, and μ is the carrier mobility. However, the MR response in such principle is very weak and thus limiting their broad application.

2.2. AMR effect

The AMR effect was initially discovered in 1857 by William Thomson [5]. It happened in the ferromagnetic materials, which arose from the spin-orbit interaction and the resistance depended on the orientation of the current relative to magnetization direction [29]. Usually, the resistance is higher for the current direction parallel to the magnetization and lower for the current perpendicular to the current. Such angle dependence could be described by the following equation:

$$R = R_0 + \Delta R \cos^2 \theta \qquad (3)$$

Apparently, the signal extrema are achieved at angles of 0° and 90°, and the steepest response slope is at an angle of 45°. Thus, to achieve the highest sensitivity, the sensor was normally designed with the initial magnetization direction to the current direction at an angle of 45°. Typical AMR response is about 1–4% [5, 29], which is good enough for allowing the use of AMR sensors in practical application.

2.3. GMR effect

GMR was discovered independently by Albert Fert and Peter Grünberg in 1988 [15, 16]. Since then, great progress in improving the GMR value has been achieved, and the use of GMR effect in practical applications has significantly changed the world. The most important application is the use of GMR read head, which has dramatically increased the areal density in HDD and brought the advent of the digital world. The significance of the GMR discovery was recognized by the Nobel Prize in Physics awarded to Fert and Grünberg in 2007. In the GMR effect, the resistance relies on the angle between magnetization directions at different locations in the materials, which could happen in granular systems [30–32] or ferro-/non-ferro-multilayer materials [33, 34]. For real application, the multilayer structure is mainly considered rather than the granular system which is normally in an uncontrollable growth condition. The basic principle of GMR effect is the spin-dependent scattering, in which a parallel direction between the current spin and magnetization can generate a low scattering, while its antiparallel direction is in a high scattering. The resistance as a function of the angle between magnetization directions is described by:

$$R(\theta) = R(0) + \Delta R[1 - \cos\theta]/2 \qquad (4)$$

where ΔR is the value of R(180°)–R(0°). This equation can show that the angle dependence of the GMR effect has a period of 360°, which forms an obvious contrast to the AMR effect with a period of 180°.

2.4. TMR effect

As one specific case of GMR, the TMR [35–37] happened as the nonmagnetic conductive layer was substituted by an insulating layer in the multilayer structure, named as magnetic tunnel junction (MTJ). In this structure, electrons can pass through this insulator by means of the quantum tunnel effect. In early reports, the insulating layer in MTJ was constituted by Al_2O_3, which demonstrated a MR level of MTJ about 40% [36]. Recently, this level has been significantly improved to 200% by using MgO as the insulating layer [17, 18]. The improvement of TMR level has significantly increased the areal density of HDD; meanwhile, it has boosted the development of spin-transfer torque-based magnetoresistive random access memory (STT-MRAM) for the next-generation memory application [38].

2.5. Geometric-enhanced MR effect

Geometric-enhanced MR is another specific case of the orbital MR associated with the material shape [24, 28]. A typical structure is consisted of a composite including conductive metal and less conductive semiconductor. The electric field, **E**, is normal to the equipotential surface of the highly conductive metal. The current density is written as $\mathbf{J} = \boldsymbol{\sigma}(H)\mathbf{E}$, where $\boldsymbol{\sigma}(H)$ is the conductivity tensor. At $H = 0$, the tensor of $\boldsymbol{\sigma}(H)$ is diagonal, so $\mathbf{J} = \sigma\mathbf{E}$ and the current flows into the conductive metal, which acts as a short circuit. At high H, the off-diagonal component of $\boldsymbol{\sigma}(H)$ gives rise to a deflection of **J** from **E**, and the deflection angle is called as Hall angle [39], which is dependent on the Hall mobility μ and H. In a sufficiently large H, the current may deflect around the conductive metal and flow in the less conductive semiconductor, which acts as an open circuit. The transition from short circuit at low H to open circuit at high H results in a geometric enhancement of the MR.

3. The research progress of MR effect in Mn_xGe_{1-x} system

3.1. MR effect in the Mn_xGe_{1-x} thin film

Mn_xGe_{1-x} could go through different MR effects through engineering Mn-doping concentration, Mn_xGe_{1-x} phase, and geometric structure. At the beginning, we pay our attention to the MR effect that happens in Mn_xGe_{1-x} thin film. The Mn_xGe_{1-x} thin film was grown on Ge substrate by a Perkin-Elmer solid-source molecular beam epitaxy (MBE) with Ge and Mn Knudsen cells. Ge substrate was cleaned by immersing in acetone and isopropyl alcohol with ultrasonic agitation, followed by dipping in 1% hydrofluoric (HF) acid. Then, the substrate was directly transferred into the MBE chamber for thin-film growth at around 200°C. After growth, its microstructure and composition were comprehensively characterized by transmission electron microscopy (TEM) equipped with energy-dispersive spectroscopy (EDS). Its magnetic property and magnetoresistance were revealed by SQUID and physical property measurement system (PPMS).

Figure 1(a) is a typical cross-sectional TEM image of the grown thin film, which shows some dark parts embedded in the Ge matrix. To understand the detailed structure, high-resolution TEM (HRTEM) was employed, and typical [110] zone axis TEM results are shown in **Figure 1(b)–(d)**.

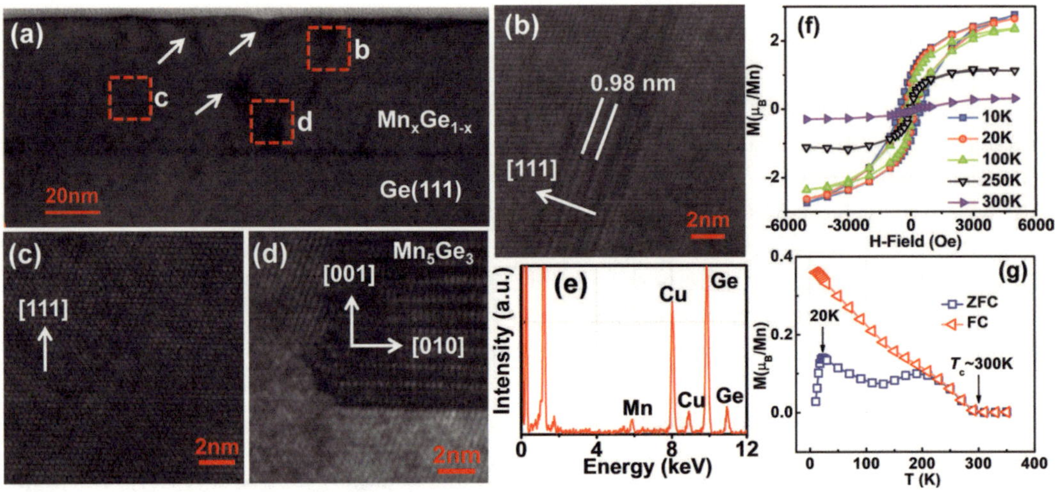

Figure 1. (a) A typical cross-sectional TEM image of the Mn_xGe_{1-x} thin film. (b)–(d) HRTEM images of the stacking faults, Mn_xGe_{1-x} DMS region, and Mn_5Ge_3 cluster in the film, respectively. (e) EDS spectrum confirming the ~4% Mn doping. (f) Temperature-dependent hysteresis loop of the Mn_xGe_{1-x} thin film. (g) ZFC and FC curves, showing a T_c around 300 K. Reproduced with permission from [26]. Copyright 2015, Elsevier.

In **Figure 1(b)**, stacking faults (SFs) could be clearly observed in the Ge matrix and feature a triplet periodicity of Ge (111) lattice spacing [40]. The formation of SFs might come from the strain accumulation of Mn doping and lattice-mismatched precipitates. **Figure 1(c)** was collected from the relatively-uniform doped region and clearly displays a very good crystallinity. Meanwhile, some precipitates could be noticeably observed in the thin-film sample, as shown in **Figure 1(d)**. It clearly shows another set of lattice structure that is different from the Ge matrix. Using the lattice spacing of the Ge as the reference, we calculated the observed lattice spaces of the cluster, which matched well with the (002) and (010) atomic planes of the hexagonal Mn_5Ge_3 phase [41, 42]. Furthermore, it can be confirmed that the Mn_5Ge_3 (002) plane was parallel to the Ge (111) plane. To determine the Mn-doping concentration, EDS was carried out, and the result reveals that the average Mn concentration is ~4%, as shown in **Figure 1(e)**. The magnetic property of the grown Mn_xGe_{1-x} film was measured by SQUID, and the result is shown in **Figure 1(f)**. Magnetic hysteresis can be observed between 10 and 250 K, and it disappears around 300 K. To further determine the T_c and detect any magnetic precipitates, zero-field-cooled (ZFC) and field-cooled (FC) magnetic measurements were performed under a small magnetic field of 200 Oe. As shown in **Figure 1(g)**, the magnetization vanishes near 300 K, indicating a T_c ~ 300 K, which further confirms the formation of Mn_5Ge_3 (T_c ~ 296 K) [43]. Two blocking temperatures coexist in the ZFC curve, with the lower one at 20 K and the higher one at 200 K, which are attributed to Mn-rich coherent Mn_xGe_{1-x} nanostructures [44] and Mn_5Ge_3 precipitates [27], respectively. Both of them could be well resolved in the TEM characterization in **Figure 1(a)**, as indicated by white arrows and red-dotted squares, respectively.

For sensor application, it should explore the MR effect of the Mn_xGe_{1-x} thin film to reveal its control parameters. To this end, the current sample accompanied by a more Mn-doped sample (6%) was fabricated into micrometer Hall bar structures by photolithography for the

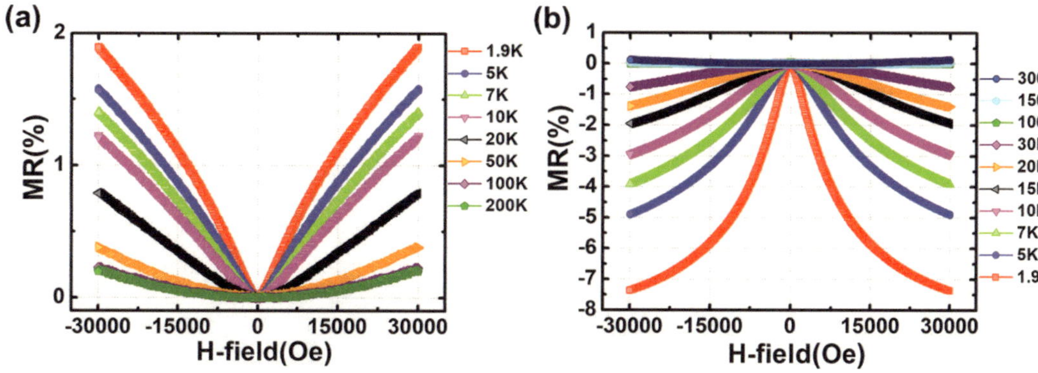

Figure 2. (a)–(b) Temperature-dependent MR of thin films with 4 and 6% Mn dopants, respectively.

magnetotransport measurement. **Figure 2** shows the temperature-dependent MR curves of the two samples with an out-of-plane magnetic field. **Figure 2(a)** is the MR curves of the 4% Mn-doped sample, and it shows only positive MR in the whole temperature range. Careful examination could find that the MR curve follows a parabolic shape in the whole temperature range except for a small deviation happening at low magnetic field at 1.9 K. The parabolic dependence exclusively indicates that the orbital MR is dominant in this sample [45], while the small deviation should come from the magnetization-enhanced orbital MR [46]. As the Mn dopants increase to 6%, however, the orbital MR is suppressed, and a negative MR is demonstrated at low temperatures, as shown in **Figure 2(b)**. It implies that the enhanced magnetization in more Mn-doped sample significantly boosts the spin-dependent scattering and thus generating the negative MR [45]. However, as the magnetization becomes weak at high temperatures, the positive MR shows up and the orbital MR is dominant again. Through tuning the Mn-doping concentration in Mn_xGe_{1-x} thin film, the engineering of negative and positive MR is conveniently realized, although it still needs to enlarge the MR value for the potential application in MR sensors.

3.2. MR effect in Mn-rich Mn_xGe_{1-x} coherent nanostructures

To precisely control the MR and seek for a large value in Mn_xGe_{1-x} system, Mn-rich Mn_xGe_{1-x} coherent nanostructures are well designed based on its growth thermal dynamics and kinetics characteristics. As already well documented, Mn atoms preferred to form intermetallic compounds [47, 48] with Ge at high growth temperatures, while they tended to form Mn-rich Mn_xGe_{1-x} coherent nanostructures [27, 44, 49, 50] at low growth temperature. By choosing proper growth conditions, Mn-rich Mn_xGe_{1-x} coherent nanostructures with different morphologies were formed by MBE, and their effect to MR properties are well discussed. Following the same cleaning procedure as described above, Ge substrate was loaded into the MBE chamber for superlattice growth. A high-quality Ge buffer layer was first deposited at 250°C, and then the growth temperature was cooled down to 70°C for the subsequent superlattice growth. Ten periods of Mn_xGe_{1-x} and Ge layers were alternatively deposited on the substrate. By adjusting the nominal thickness of Ge space layer from 6 to 25 nm while keeping the Mn_xGe_{1-x} layer at

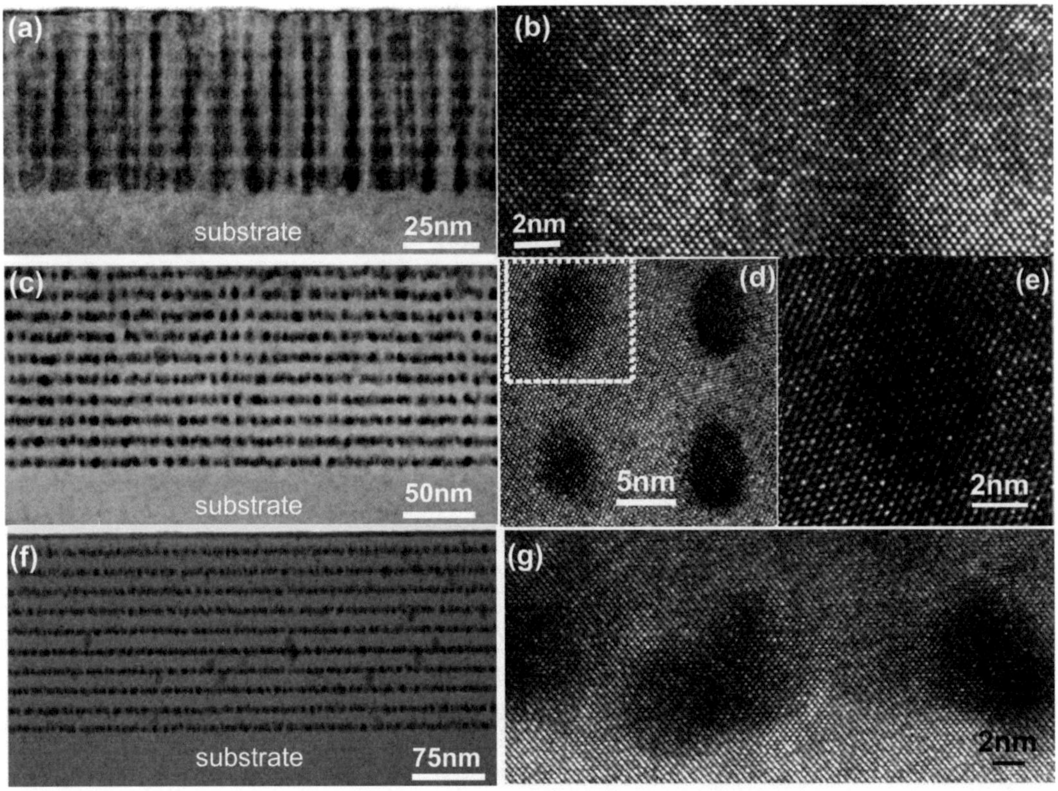

Figure 3. (a) A typical cross-sectional TEM image of the Mn_xGe_{1-x} nanocolumns. (b) Its HRTEM image, showing the coherent Mn_xGe_{1-x} nanocolumns. (c) A typical TEM image of the Mn_xGe_{1-x} nanodots. (d) Its HRTEM image, showing well vertically aligned Mn_xGe_{1-x} nanodots. (e) The zoom-in HRTEM image, showing its diamond lattice structure. (f) A typical TEM image of the nanowell structure. (g) HRTEM image of a single-layer nanowell consisting of coherent MnGe nanodots.

about 4 nm, Mn_xGe_{1-x} nanocolumns, nanodots, and nanowells were obtained, respectively. **Figure 3(a)** shows a typical cross-sectional TEM image of the Mn_xGe_{1-x} nanocolumns with a nominal Ge space layer of 6 nm, in which well-aligned nanocolumns with dark contrast can be clearly observed. To reveal the detailed lattice structure, HRTEM experiments were carried out, and the result is shown in **Figure 3(b)**. Careful examination of the HRTEM image verifies that the Mn_xGe_{1-x} nanocolumns have the same diamond lattice structure as the Ge matrix, showing a coherent growth. The formation of nanocolumns rather than layer structure indicates that Mn atoms not only agglomerate laterally but also migrate vertically into the adjacent Ge space layers [51]. After increasing the Ge space layer thickness to 11 nm, the superlattice evolves from a nanocolumn structure to a nanodot structure. From the cross-sectional TEM image shown in **Figure 3(c)**, 10 periods of nanodots with Ge space layers are clearly observed, in which the nanodots are well aligned along the vertical direction. The alignment may follow the fact that the buried MnGe nanodots induce an elastic strain in the thin Ge space layer, which provides a preferential nucleation site for the formation of new MnGe nanodots. Through further HRTEM characterizations, these nanodots show relatively uniform size distribution with an elliptical shape (dimension of 5.5 ± 0.5 and 8 ± 0.3 nm in the horizontal and vertical directions, respectively), as demonstrated in **Figure 3(d)**. The fact that the vertical diameter of the nanodot is much larger than the thickness of the MnGe layer again suggests that Mn atoms migrate vertically into the adjacent Ge space layer. **Figure 3(e)** clearly shows the coherent growth of MnGe nanodot in the Ge matrix. After further increasing the Ge space layer thickness to 25 nm, MnGe nanowells with 10 periods were obtained, and a typical TEM image is shown in **Figure 3(f)**. Noticeably, the nanowells are also composed of dense MnGe nanodots, whereas the nanodots are not aligned in the vertical direction. It may be due to the strain release on the top surface of the thick Ge space layer, and hence there are no energy-preferable positions for subsequent nucleation of the MnGe nanodots [52]. Similar to the previous two cases, the MnGe nanodots inside the nanowells are also coherent with the surrounding Ge matrix with a diameter range of 4–10 nm, as shown in **Figure 3(g)**.

The magnetic properties of the formed nanostructures are disclosed by SQUID. Due to the similarity, we took the case of the Mn_xGe_{1-x} nanocolumn as an example, and the result is shown in **Figure 4**. **Figure 4(a)** shows the temperature-dependent M-H curves, which clearly demonstrate the ferromagnetism from 10 to 175 K and the paramagnetism at 300 K. At 10 K, the film exhibits a saturation magnetic moment of 104 kAm^{-1} that is estimated to be 0.24 μ_B per Mn atom. This gives a fraction of 8% of Mn activated in Mn_xGe_{1-x} when considering the value of 3 μ_B in each fully active Mn atom [53]. To further understand its magnetic property, the ZFC and FC curves were performed with a magnetic field of 200 Oe in SQUID, and a typical result is shown in **Figure 4(b)**. The differences between them give an insight into the anisotropic barrier distribution, blocking temperature, and Curie temperature [54]. Two blocking temperatures could be well resolved with one at 25 K and the other at 250 K. As discussed above, the two blocking temperatures indicate the coexistence of Mn-rich Mn_xGe_{1-x} nanostructures and Mn_5Ge_3 nanoparticles in the film. The former has been well recognized in **Figure 3**, whereas the latter could be occasionally observed by comprehensive TEM characterization. Such fact proves that the ZFC and FC measurement in SQUID is more sensitive to detect magnetic particles than TEM [44, 54]. The Curie temperature of the nanocolumns is around 300 K proven from the almost zero magnetic moment at this temperature, which further confirms the existence of Mn_5Ge_3 (T_c ~ 296 K).

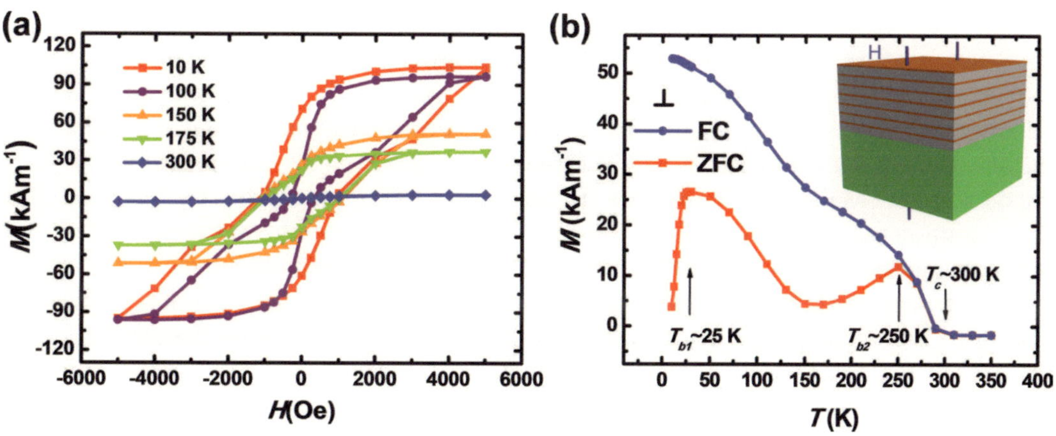

Figure 4. (a) Temperature-dependent hysteresis loops of Mn_xGe_{1-x} nanocolumns with out-of-plane external magnetic field. (b) ZFC and FC curves. Inset is a schematic drawing of the sample setup during the SQUID measurements.

Since the Mn-rich Mn_xGe_{1-x} coherent nanostructures demonstrate different morphologies, it is of great interest to reveal their MR properties and explore the potential application for MR sensors. The samples were fabricated into micrometer-size Hall bar structures, and the measurements were performed in PPMS. The temperature range is from 2 to 300 K with an external magnetic field up to 10 T in an out-of-plane direction. For Mn_xGe_{1-x} nanocolumns, negative MR is observed below 50 K, and a transition to positive MR happens at high temperatures, as shown in **Figure 5(a)**. The origin of the negative MR should come from the spin-dependent scattering mechanism. In the absence of magnetic field, the carrier transport between the nanocolumns with relatively random spin alignment is believed to result in a strong spin-dependent scattering and thus in a high resistance. The high magnetic field would align the spin of the nanocolumns preferentially in one direction. The reduction of spin scattering leads to a low resistance, hence generating a negative MR. The positive MR at high temperatures follows a parabolic shape, which indicates that an orbital MR effect is dominant.

However, the negative MR disappears in Mn_xGe_{1-x} nanowells and nanodot structures, respectively, shown in **Figure 5(b)** and **(c)**, and instead only positive MR presents in the entire temperature range. Careful examination can find that the positive MR in both of the samples does not follow a perfectly parabolic shape and shows very large values at low temperatures. Especially, the MR value in Mn_xGe_{1-x} nanodot structure is as high as 1000% at 2 K, which is hard to be simply explained by orbital MR. Instead, the geometric-enhanced MR due to the existence of conductive Mn_xGe_{1-x} dots is likely to respond to the large MR [24, 43]. To elucidate the underlying physics of the geometric effect, we consider the current density and the total electric field in semiconductors, which can be described by $\mathbf{J} = \boldsymbol{\sigma}(H)\mathbf{E}$, where the $\boldsymbol{\sigma}(H)$ is described as follows:

$$\sigma(H) = \begin{pmatrix} \frac{\sigma}{1+\beta^2} & \frac{\sigma\beta^2}{1+\beta^2} & 0 \\ -\frac{\sigma\beta^2}{1+\beta^2} & \frac{\sigma}{1+\beta^2} & 0 \\ 0 & 0 & \sigma \end{pmatrix} \tag{5}$$

Here, $\beta = \mu H$. In the absence of magnetic field, the tensor $\boldsymbol{\sigma}(H)$ is diagonal, and the current density can be simply described by $\mathbf{J} = \sigma\mathbf{E}$. In this scenario, the current flowing is parallel to the electric field and concentrated into the conductive Mn_xGe_{1-x} nanodots in the Mn_xGe_{1-x} nanowell or nanodot structures, acted as a "short circuit." However, at a high magnetic field, the off-diagonal term of $\boldsymbol{\sigma}(H)$ indicates that the current is deflected from conductive Mn_xGe_{1-x} nanodots to the Mn-dilute Ge matrix with low conductivity, resembling an "open circuit." The transition from the "short circuit" at zero fields to the "open circuit" at high fields produces an increase in resistance. The increased amplitude is significantly dependent on the resistance ratio of Mn_xGe_{1-x} nanodots to the Mn-dilute Ge matrix, and the deflection angle is defined by magnetic field and carrier mobility [24]. Through this mechanism, it is also very simple to explain why the Mn_xGe_{1-x} nanodot structure shows much larger MR compared to Mn_xGe_{1-x} nanowell structure. Compared to Mn_xGe_{1-x} nanodot structure, the disordered conductive nanodots in Mn_xGe_{1-x} nanowell structure make it hard to bypass the conductive nanodots even after deflection around one dot, thus generating lower MR ratio.

http://dx.doi.org/10.5772/intechopen.70206

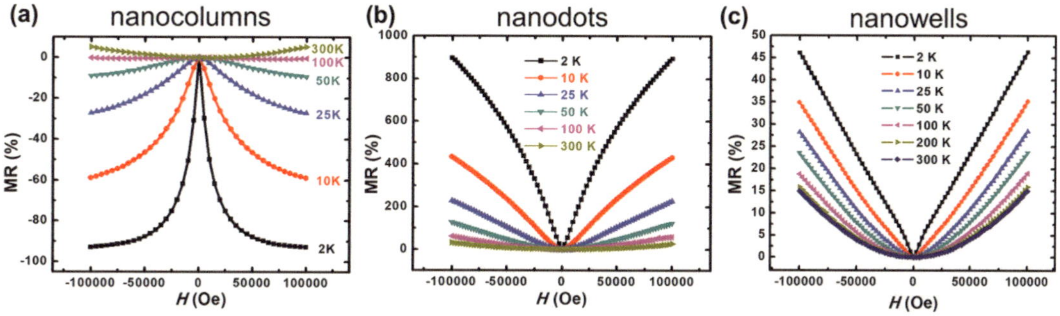

Figure 5. (a)–(c) Temperature-dependent MR curves of the Mn_xGe_{1-x} nanocolumns, nanodots, and nanowells, respectively. The Mn_xGe_{1-x} nanocolumns show negative MR below 50 K, while a positive value dominates at high temperature. For Mn_xGe_{1-x} nanodots and nanowells, however, only positive MR is observed in the whole temperature range from 2 to 300 K.

The above results clearly demonstrate that the negative-to-positive MR and even large MR can be well engineered through well designing the Mn-rich Mn_xGe_{1-x} coherent nanostructures. Therefore, it points out a new direction to easily engineer Mn_xGe_{1-x} nanostructures through strain approach, which may provide a great advantage in designing MR sensors for more functionality.

3.3. Geometric-enhanced and electric-field controlled MR in Mn_xGe_{1-x} nanomesh

The strain engineering of Mn_xGe_{1-x} nanostructures provides a potential approach for satisfying the demand of MR sensors with multifunctionality; however, the self-assembly formation of Mn_xGe_{1-x} nanodots increases the difficulty in accurate control of MR. Therefore, we pay our attention to the pattern-assistant growth of Mn_xGe_{1-x} nanostructures and disclose their MR property. In this section, Mn_xGe_{1-x} nanomesh is demonstrated, which could simultaneously provide the nanostructure benefit [55] and large-scale uniform fabrication [28]. The growth of Mn_xGe_{1-x} nanomesh is also proceeded in the MBE chamber. Before that, great effort has been devoted to fabricate the pattern structure. A 100-nm-thick SiO_2 thin film was firstly deposited on a Ge (111) wafer by PECVD, followed by the formation of a large-scale and close-packed hexagonal single layer of nanospheres on the SiO_2 substrate, as shown in **Figure 6(a)**. By adjusting O_2-plasma etching time, the nanospheres were successfully shrunk to 160 nm with a 60 nm gap between them as shown in **Figure 6(b)**. Using the nanosphere as the mask, the pattern was transferred to the bottom SiO_2 layer by a two-step etching. Dry etching was firstly employed to etch SiO_2 layer till a 10 nm SiO_2 left. Then, wet etching was hired to remove the left SiO_2 layer. After dissolving the nanospheres, periodic SiO_2 nanopillars were obtained on the substrate, and a typical scanning electron microscopy (SEM) image is shown in **Figure 6(c)**. After carefully cleaning, the substrate was loaded into the MBE chamber for the Mn_xGe_{1-x} growth. After degassing at 600°C for 30 min, the patterned substrate was in situ cooled down to 160°C for the Mn_xGe_{1-x} nanomesh growth with a Ge growth rate of 0.2 Å/s and a controlled Mn flux as the dopant source.

The SiO_2 mask was subsequently removed after MBE growth by selective etching, and only Mn_xGe_{1-x} nanomesh was remained. A typical morphology of the sample was captured by SEM

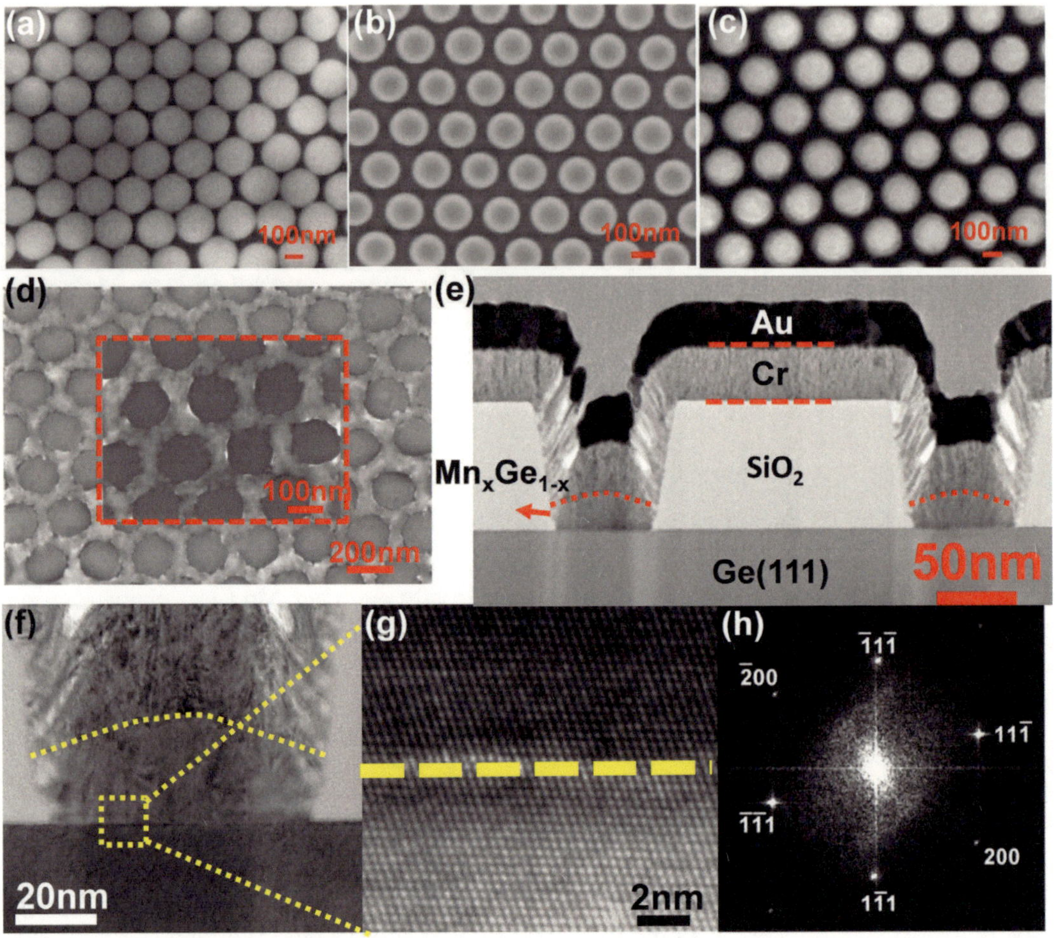

Figure 6. (a) Self-assembly growth of close-packed single layer of nanospheres on the Ge substrate. (b) O_2-plasma etching of the nanospheres to reach the desired size. (c) SiO_2 nanopillars formed by dry etching, masked by the nanospheres. (d) Typical SEM image of the Mn_xGe_{1-x} nanomesh, with a nanomesh width of 60 nm and a nanohole diameter of 160 nm. The inset is the magnified SEM image. (e) Cross-sectional TEM image in a low resolution, showing the Mn_xGe_{1-x} nanomeshes defined well by SiO_2 mask. (f) Magnified cross-sectional TEM image of the Mn_xGe_{1-x} nanomesh. (g) HRTEM image of the interface between the nanomesh and substrate, clearly showing a perfect coherent growth. (h) Its Fourier transform image.

as shown in **Figure 6(d)**. A periodic nanomesh structure exhibits a nanomesh width of 60 nm and a hole diameter of 160 nm; a magnified SEM image is also shown in the inset. To further characterize the microstructure of the formed nanomesh, cross-sectional TEM was employed, and the results are shown in **Figure 6(e)–(h)**. The focused ion beam was employed to cut the sample along the diameter of the hole. Before that, a Cr/Au layer was deposited to protect the sample from damage of the ion beam. **Figure 6(e)** is a low-resolution cross-sectional TEM image of the Mn_xGe_{1-x} nanomesh, which clearly shows that the nanomesh is grown on the Ge substrate as defined by the SiO_2 pattern. The zoom-in image shows that the Mn_xGe_{1-x} nanomesh has a height of 25 nm and a width of 60 nm, consistent with the SEM result. HRTEM image clearly demonstrates the coherent growth of the Mn_xGe_{1-x} nanomesh on Ge substrate, as shown in **Figure 6(f)**. It does not reveal any observable precipitate. The Fourier transform image as shown in **Figure 6(h)** gives only one set of periodic patterns, indicating a perfect epitaxial growth. This image can be indexed to the [011] zone axis of the Ge diamond lattice, and the epitaxial growth direction is along [111].

SQUID measurement was performed in the following to well understand the magnetic property of the Mn_xGe_{1-x} nanomesh. **Figure 7(a)** shows the temperature-dependent hysteresis loops of the sample, when an external field is applied parallel to the sample surface. The S-shaped hysteresis loops indicate the ferromagnetism above 350 K. **Figure 7(b)** is the magnified hysteresis loop obtained at 10 K, clearly showing a small coercivity of 100 Oe with a saturation magnetization of 0.87 μ_B per Mn. At 350 K, the magnified hysteresis loop demonstrates that the coercivity still remains a value of 40 Oe, as shown in **Figure 7(c)**. Furthermore, Arrott plots [55] were also used to evaluate the T_c, as shown in **Figure 7(d)**. We observe that even at 350 K, the intercept, namely, reciprocal of the susceptibility, does not vanish, indicating that the T_c has not been reached yet. The extrapolated dashed line indicates that the T_c is beyond 350 K, which agrees well with the hysteresis loops. **Figure 7(e)** shows the temperature-dependent M_s ranging from 10 to 400 K, and it clearly shows a weak temperature dependence and a large magnetization remaining at 400 K. All of the data support that the T_c is over 400 K. The temperature-dependent coercivity is shown in **Figure 7(f)**, demonstrating a coercivity decrease from 100 to 35 Oe in the temperature range from 10 to 400 K. The small coercivity indicates the soft ferromagnetism of our sample, which may come from the Mn-dilute nature.

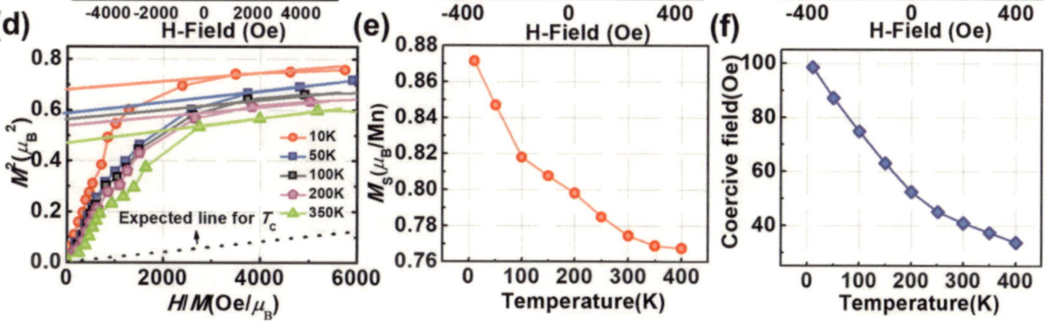

Figure 7. (a) Magnetic hysteresis loops of the Mn_xGe_{1-x} nanomeshes measured at different temperatures from 10 to 350 K. (b)–(c) The magnified hysteresis loop obtained at 10 and 350 K, respectively. (d) Arrott's plots showing that the T_c is above 350 K. (e)–(f) The temperature-dependent saturation moment and coercivity, respectively.

In this unique nanostructure, it is of great interest to investigate its MR property. The sample is fabricated into a micrometer-size Hall bar structure, and the measurement is performed in PPMS, as shown in **Figure 8. Figure 8(a)** shows the temperature-dependent resistivity of the Mn_xGe_{1-x} nanomesh under the magnetic field of 0 T (blue square) and 4 T (red circle), respectively, demonstrating a huge difference. A clear metal-to-insulator transition without applying magnetic field (0 T) can be observed with a low-temperature (T < 30 K) activation region and a high-temperature (T > 30 K) saturation region. Form the Arrhenius relation [56], the activation energy of the nanomesh is estimated to be about 11 meV, which is lower than the substitutional Mn acceptor energy level (160 meV). The underlying mechanism comes from the high-doping level and the presence of exchange interaction, inducing the boarding and possible splitting of the impurity band. Above 30 K, the R-T curve could be well fitted by a power-law relation (T^α) with $\alpha \approx 1.6$, close to the value 1.5 predicted for hole scattering by phonons in Ge. The fitting was plotted in red over the blue data in the R-T curve. An intriguing phenomenon is the observation of a giant positive MR in the nanomesh with an out-of-plane magnetic field, as shown in **Figure 8(b)–(d)**. At a magnetic field of 4 T,

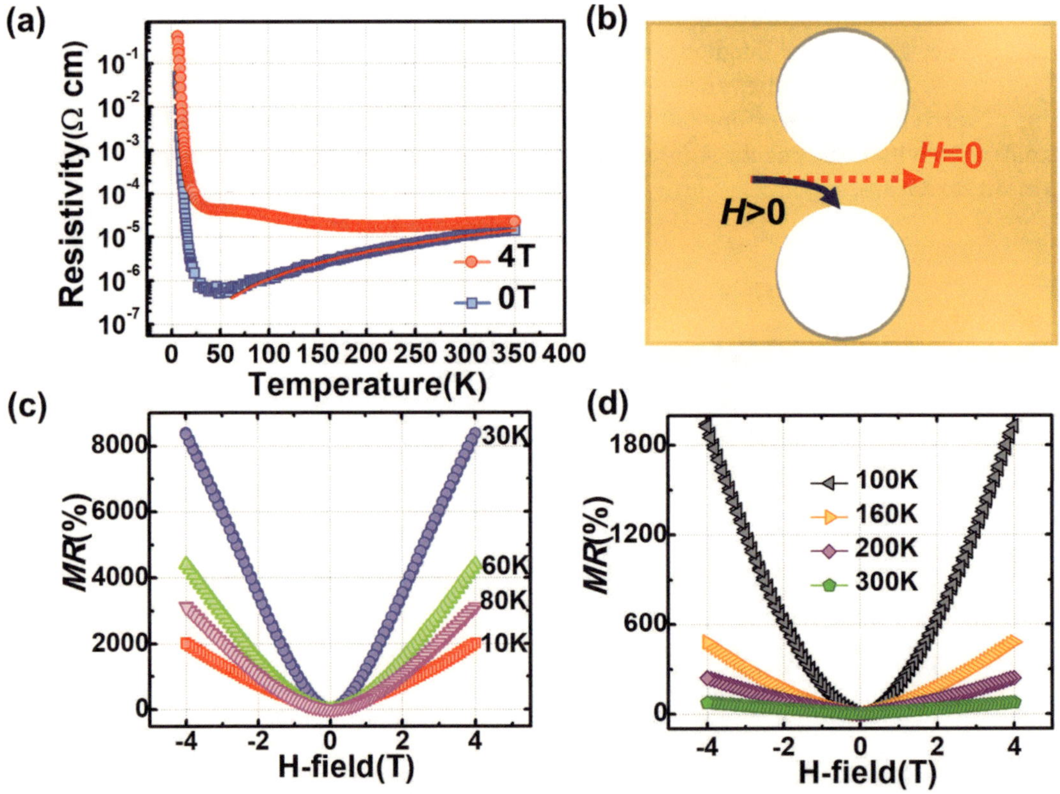

Figure 8. (a) Temperature-dependent resistivity of the Mn_xGe_{1-x} nanomeshes measured without an applied magnetic field (square symbol) and with a magnetic field of 4 T (circle symbol), clearly showing a metal-to-insulator transition. The solid line is the fitting curve. (b) The schematic illustration of scattering mechanism. (c)–(d) Temperature-dependent MR measured at low-temperature and high-temperature regions, respectively.

the MR as high as 2000% at 10 K with the maximum of 8000% at 30 K is observed and still remains 75% at 300 K. Analogously, the giant MR could not be simply attributed to the orbital MR effect, which only contributes a small value. Instead, the geometric-enhanced MR is very plausible to explain the result. To understand this phenomenon, the nanomesh structure could be considered as a highly conductive percolation network with periodic nanoholes. Without applying the magnetic field, the current flows through the Mn_xGe_{1-x} nanomesh with the current directions parallel to the local electric field. As the magnetic field is applied, the current is deflected due to the Lorentz force; the current and local electric fields are no longer collinear. The angle between them is determined by the Hall angle $\theta = \arctan(\mu_H H)$, where the μ_H is the Hall mobility. For a sufficiently high magnetic field, the current is obviously deflected from the highly conductive nanomesh to the insulated nanoholes, resulting in a high resistance. The transition from the extremely low resistance at zero magnetic field to the extremely high resistance at a large magnetic field gives rise to the giant MR, as illustrated in **Figure 8(b)**. Thus, it may be concluded that the lower the initial resistance of the nanomesh is, the larger the MR became at a given magnetic field. It can be verified from the deviation of the R-T curves of 4 T and 0 T, as already shown in **Figure 8(a)**. The largest deviation happens at the lowest resistivity (at 30 K and 0 T), which agrees well with our proposed model.

The Mn_xGe_{1-x} nanomesh is a type of diluted magnetic semiconductor [53, 57] that could provide the ability of electric-field control of ferromagnetism. Therefore, the electric-field controlled MR could be possibly realized to develop new MR sensors with more functionalities. To demonstrate the spin-related MR effect, it is in need of weakening the geometric-enhanced MR effect. One effective approach is to increase the initial resistance of the nanomesh while reducing the carrier mobility. To this end, more Mn-doped Mn_xGe_{1-x} nanomesh was grown, and the transport property is shown in **Figure 9**. **Figure 9(a)** shows the R-T curve of the nanomesh, in which a metal-to-insulator transition is observed; however, the resistivity was found to be more than one order of magnitude larger than the above sample. The increased resistivity may come from the increased scattering center [58]. **Figure 9(b)** and **(c)** shows the temperature-dependent MR curves of the sample. Intriguingly, the geometric-enhanced MR becomes less pronounced, which may be due to the dramatically increased resistivity and the decreased Hall angle from the lower mobility. Instead, a negative MR for temperature below 40 K and a positive MR above 160 K are observed. In the intermediate temperature region (40–160 K), the MR contains two contributions: a positive MR appears at a low magnetic field and a negative slope at high field. Such negative-to-positive MR transition could be attributed to two competitive effects: the spin-dependent scattering by magnetic polarons gave rise to a negative MR [59, 60], and their spatial fluctuations led to the positive MR [59, 60]. As a magnetic field is applied, the carrier mobility increases due to the suppression of the spin-dependent scattering by the magnetic polarons, giving rise to the negative MR, which is proportional to the susceptibility ($\chi(H)$) of the sample. However, the spin-dependent scattering effect by the magnetic polarons decreases with increasing temperature, which makes the positive MR from the spatial fluctuations of magnetic polarons gradually to show up. Due to the strong *p-d* exchange coupling in Mn_xGe_{1-x}, the magnetic polarons will cause a strongly localized valence band splitting into two spin subbands $\varepsilon = \pm 1/2SJM(r)$ [61], where J is the

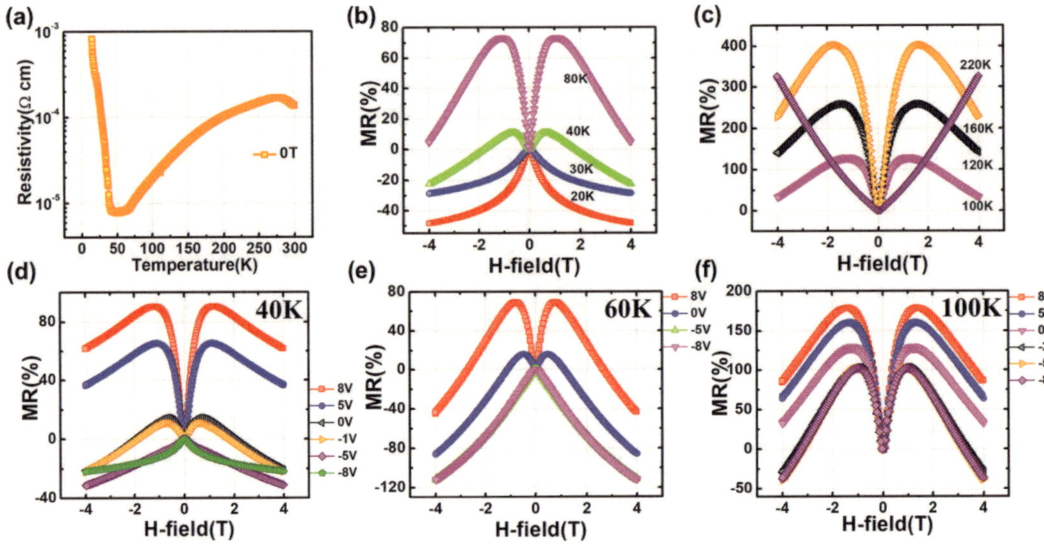

Figure 9. (a) R-T curve of the Mn_xGe_{1-x} nanomeshes with more Mn dopants. (b)–(c) Temperature-dependent MR measured at low-temperature (<80 K) and high-temperature (>100 K) regions, respectively, showing a MR transition from a negative to positive value. (d) Electric-field controlled MR at 40 K, clearly showing a transition from negative to positive when the gate was biased from negative to positive. (e)–(f) Electric-field controlled MR transition at 60 and 100 K, respectively.

exchange coupling energy, S is the spin of electron, and M(r) is the local magnetic moment of the magnetic polaron. This could form a complicated landscape for the valence band with hills and valleys, serving as hole traps. With increasing magnetic field, the hole traps become deeper, the number of itinerant holes decreases, and thus the MR increases. This positive MR is proportional to the square of the magnetization ($M^2(H)$) [59], which can be reflected from the steep slope of the positive MR. As the temperature further increases, the contribution from the spin-dependent scattering becomes weaker, and finally only positive MR dominates the transport.

Furthermore, electric-field control of MR was measured under different gate biases. For that, a 25-nm-thick Al_2O_3 layer was deposited by atomic layer deposition (ALD) on the nanomesh surface as the gate dielectric, followed by the e-beam evaporation of Cr/Au as the gate metal contact. Due to the unique nanomesh structure, the almost-wrap-around gate can be realized to provide a 3D electric-field control of the conduction channel, thus giving a highly efficient and robust carrier modulation. **Figure 9(d)–(f)** shows the gate bias-dependent MR of the Mn_xGe_{1-x} nanomesh sample measured at 40, 60, and 100 K, respectively. When the gate bias changes from −8 to 8 V, a clear negative-to-positive MR transition is observed at 40 K. As described above, this phenomenon should also stem from the competitive effect between the spin-dependent scattering-induced negative MR and the spatial fluctuation-induced positive MR. As already reported [53], the hole-mediated ferromagnetism in Mn-doped Ge enables the electric-field control of magnetic phase transition from a strong ferromagnetism to a soft ferromagnetism, when changing the gate bias from negative to positive. Therefore, as a negative gate bias is applied on our sample, the enhanced ferromagnetism with larger susceptibility ($\chi(H)$) in the Mn_xGe_{1-x} nanomesh could provide a strong spin-dependent scattering, consequently giving rise to a negative MR. When the gate bias is swept to positive value, the weakened ferromagnetism gives rise to a weaker spin-dependent scattering. From another point of view, as the gate bias goes to the positive range, the decreased density of magnetic polarons due to reduced carrier density probably transforms the system from an overlap-together state to a disconnecting state, which thus increases the polaron fluctuation, and hence the positive MR dominates. In addition, gate-dependent MR curves at 60 K and 100 K are shown in **Figure 9(e)** and **(f)**, respectively. At 60 K, the MR transition from the negative to positive value can still be clearly observed when tuning the gate bias. However, the control effect is not as strong as that of 40 K in the negative bias range. This effect can be explained by the fact that even at 0 V, the hole density at 60 K is already high enough to align the magnetic moments of most Mn ions along one direction. Further increasing the negative bias does not significantly enhance the ferromagnetism. At 100 K, it shows a similar effect, i.e., an even weaker control effect in the negative bias range. This extraordinary property may offer a great advantage for designing new MR sensors with voltage-controlled, low-power, high-sensitivity, and nonvolatile functionalities.

4. The potential application of Mn_xGe_{1-x} in MR sensors

The development and utilization of MR sensors have tremendously impacted human beings' life, which refers to a broad range of aspects, including magnetic HDD [4, 23], electric compass [3], biomedical sensors [8, 9], car traffic monitoring [7], and antitheft system [5], among others. The future development in MR sensors should rely on these two perspectives: applications and physics. The discovery of new physics will lead to a new sensor technology toward much smaller size, lower power consumption, lower cost, and higher performance, thus broadening the applications. In turn, the economic benefit from the application will promote further developing high-performance sensors with new physics. In light of that, Mn_xGe_{1-x}-based MR sensors with high compatibility with silicon technology should have a high potential to satisfy this demand. Here, we will give a discussion for its potential application in different functionalized MR sensors.

4.1. High-density information storage

The density of information storage in HDD has significantly increased in the past two decades, which promotes the coming of cloud computing age. The advent of MR read head technology transitioning from the AMR [19] read head to the GMR [34] and TMR [37] read head is the major drive in remarkably increasing storage density of HDD. The GMR and TMR sensors in read head usually comprise two magnetic metal layers separated by a thin nonmagnetic space layer. The basic operation of the magnetoresistance head is to convert the magnetic field that exists above the data bits recorded on the disk to a change in resistance that can be read out. Therefore, the MR sensitivity is extremely important and should be high enough to distinguish two states (0 and 1); meanwhile, the material for MR sensor could be easily scaled down and integrated with other electric components. Compared to the metal component in current MR sensor, Mn_xGe_{1-x} could have high compatibility with the mature Si technology [54] and hence significantly reduces the manufacturing cost. By utilizing the well-designed giant MR in Mn_xGe_{1-x} system, it will be of much interest to develop Mn_xGe_{1-x}-based MR sensor for high-density magnetic record technology.

4.2. Biomedical sensor

The medical industry has been always striving for noninvasive methods for diagnosing human illness. Due to the electric nature of brain and neuron activity, magnetic sensing is an effective way to detect the brain illness. Meanwhile, it can also be used for unraveling the mystery of the brain how to functionalize, which is of great help for building future brain computer [62] with extremely low power consumption and high performance. Besides, other parts in the body without electric activity can also be detected by magnetic sensor. One effective way is to decorate the cell with nonharmful magnetic particles as the markers. For instance, Wang et al. reported a CMOS magnetic sensor to monitor the pulsatile movements of cardiac progenitor cells tagged with magnetic particles [8]. In addition, magnet tracker can also be used to determine the position of the medical tool inside the body to observe biomechanical motions [3]. In fact, the magnetic signal in biological system normally is very small and usually needs big and heavy equipment for detecting. Therefore, highly sensitive MR sensor is necessary toward miniaturizing the equipment. Mn_xGe_{1-x} appears to be an ideal candidate for fabricating biomedical MR sensor with high sensitivity, high compatibility with CMOS technology, and no poison to human body.

5. Conclusion

MR sensors have dramatically changed the life of human beings in a wide range of applications, including automotive sensor, traffic monitor, mobile phone, HDD, and biomedical sensors, among others. The MR sensors are being developed toward much lower cost, lower power consumption, higher compatibility with CMOS technology, and higher sensitivity. Addressing such demand calls for new material candidate with new physics, improved sensitivity, and easy manufacturing. In this context, we give a review in the recent progress in Mn_xGe_{1-x} system, including material growth and magnetic and magnetotransport properties. The Mn_xGe_{1-x} evolved from thin-film superlattice to patterned nanostructures, and their MR behavior could be well engineered and transformed between the negative and positive. Furthermore, geometric-enhanced giant MR as high as 8000% at 30 K at 4 T is demonstrated. More intriguingly, the electric-field controlled MR emerges, which not only demonstrates great physical meaning but also significantly enhances the functionality of MR sensors with more tuning dimensions. The Mn_xGe_{1-x} system with the advantages of well-controlled MR, and high compatibility with Si technology may set a stage for designing a new breed of MR sensors applicable in magnetic read head and biomedical sensor with much higher performance.

Acknowledgements

This work is supported by the National Natural Science Foundation of China under Grant Nos. 11644004 and 61627813 and the International Collaboration Project B16001. We also gratefully acknowledge the FAME Center, one of the six centers of STARnet, a Semiconductor Research Corporation (SRC) program sponsored by MARCO and DARPA.

Author details

Tianxiao Nie[1,2]*, Weisheng Zhao[1] and Kang L. Wang[2]

*Address all correspondence to: nietianxiao@gmail.com

1 Fert Beijing Institute, BDBC, and School of Electronic and Information Engineering, Beihang University, Beijing, China

2 Device Research Laboratory, Department of Electrical Engineering, University of California, Los Angeles, California, USA

References

[1] Clarke J, Koch RH. The impact of high-temperature superconductivity on SQUID magnetometers. Science. 1988;**242**:217-223

[2] Meydan T. Application of amorphous materials to sensors. Journal of Magnetism & Magnetic Materials. 1994;**133**:525-532

[3] Ripka P. Magnetic Sensors and Magnetometers. Vol. 13; 2002

[4] Fullerton EE, Childress JR. Spintronics, magnetoresistive heads, and the emergence of the digital world. Proceedings of the IEEE. 2016;**104**:1787-1795

[5] Lenssen KMH, Adelerhof DJ, Gassen HJ, Kuiper AET, Somers GHJ, Zon JBADV. Robust giant magnetoresistance sensors. Sensors & Actuators A Physical. 2000;**85**:1-8

[6] Lenz JE. A review of magnetic sensors. Proceedings of the IEEE. 1990;**78**:973-989

[7] Ryoo Y-J, Park J-H. Design and development of magnetic position sensor for magnetic guidance system of automated ground vehicle. In: 2012 12th International Conference on Control, Automation and Systems (ICCAS): IEEE; 2012. pp. 988-991

[8] Wang H, Mahdavi A, Tirrell DA, Hajimiri A. A magnetic cell-based sensor. Lab on a Chip. 2012;**12**:4465-4471

[9] Baselt DR, Lee GU, Natesan M, Metzger SW, Sheehan PE, Colton RJ. A biosensor based on magnetoresistance technology. Biosensors & Bioelectronics. 1998;**13**:731-739

[10] Mujika M, Arana S, Castaño E, Tijero M, Vilares R, Ruano-López JM, Cruz A, Sainz L, Berganza J. Magnetoresistive immunosensor for the detection of *Escherichia coli* O157:H7 including a microfluidic network. Biosensors & Bioelectronics. 2009;**24**:1253

[11] Valenzuela R, Vazquez M, Hernando A. A position sensor based on magnetoimpedance. Journal of Applied Physics. 1996;**79**:6549-6551

[12] Huang WS, Lu CC, Jeng JT. A novel 3D CMOS micro-fluxgate magnetic sensor for low magnetic field detection. In: 2010 IEEE Sensors; 2010. Pp. 1791-1794

[13] Reig C, Mukhopadhyay SC, Cardoso S. Giant magnetoresistance (GMR) sensors. Vol. 6; 2013

[14] Heidari H, Bonizzoni E, Gatti U, Maloberti F. A CMOS current-mode magnetic hall sensor with integrated front-end. IEEE Transactions on Circuits and Systems I: Regular Papers. 2015;**62**:1270-1278

[15] Baibich MN, Broto JM, Fert A, Van Dau FN, Petroff F, Etienne P, Creuzet G, Friederich A, Chazelas J. Giant magnetoresistance of (001)Fe/(001)Cr magnetic superlattices. Physical Review Letters. 1988;**61**:2472-2475

[16] Binasch G, Grünberg P, Saurenbach F, Zinn W. Enhanced magnetoresistance in layered magnetic structures with antiferromagnetic interlayer exchange. Physical Review B. 1989;**39**:4828-4830

[17] Ferreira R, Wisniowski P, Freitas PP, Langer J, Ocker B, Maass W. Tuning of MgO barrier magnetic tunnel junction bias current for picotesla magnetic field detection. Journal of Applied Physics. 2006;**99**:08K706-708K709

[18] Ikeda S, Hayakawa J, Ashizawa Y, Lee YM, Miura K, Hasegawa H, Tsunoda M, Matsukura F, Ohno H. Tunnel magnetoresistance of 604% at 300K by suppression of Ta diffusion in CoFeB/MgO/CoFeB pseudo-spin-valves annealed at high temperature. Applied Physics Letters. 2008;**93**:082508-082510

[19] Tsang C, Chen MM, Yogi T, Ju K. Gigabit density recording using dual-element MR/inductive heads on thin-film disks. In: Magnetics Conference, 1990. Digests of Intermag '90. International, p CA; 1990

[20] Heim DE, Fontana RE, Jr, Tsang C, Speriosu VS. Design and operation of spin valve sensors. IEEE Transactions on Magnetics. 1994;**30**:316-321

[21] Maat S, Marley AC. Physics and Design of Hard Disk Drive Magnetic Recording Read Heads. Netherlands: Springer; 2016

[22] Charap SH, Lu PL, He Y. Thermal stability of recorded information at high densities. IEEE Transactions on Magnetics. 1997;**33**:978-983

[23] Moser A, Takano K, Margulies DT, Albrecht M, Sonobe Y, Ikeda Y, Sun S, Fullerton EE. Magnetic recording: Advancing into the future. Journal of Physics D: Applied Physics. 2002;**35**:R157-R167

[24] Solin SA, Thio T, Hines DR, Heremans JJ. Enhanced room-temperature geometric magnetoresistance in inhomogeneous narrow-gap semiconductors. Science. 2000;**289**:1530-1532

[25] Thio T, Solin SA. Giant magnetoresistance enhancement in inhomogeneous semiconductors. Applied Physics Letters. 1998;**72**:3497-3499

[26] Nie T, Tang J, Wang KL. Quest for high-Curie temperature Mn_xGe_{1-x} diluted magnetic semiconductors for room-temperature spintronics applications. Journal of Crystal Growth. 2015;**425**:279-282

[27] Faxian X, Yong W, Kin W, Yi Z, Xufeng K, Jin Z, Wang KL. MnGe magnetic nanocolumns and nanowells. Nanotechnology. 2010;**21**:255602-255606

[28] Nie T, Tang J, Kou X, Gen Y, Lee S, Zhu X, He Q, Chang L-T, Murata K, Fan Y, Wang KL. Enhancing electric-field control of ferromagnetism through nanoscale engineering of high-Tc Mn_xGe_{1-x} nanomesh. Nature Communications. 2016;**7**:12866

[29] Fert A, Campbell IA. Electrical resistivity of ferromagnetic nickel and iron based alloys. Journal of Physics F Metal Physics. 2001;**6**:849-871

[30] Berkowitz AE, Mitchell JR, Carey MJ, Young AP, Zhang S, Spada FE, Parker FT, Hutten A, Thomas G. Giant magnetoresistance in heterogeneous Cu-Co alloys. Physical Review Letters. 1992;**68**:3745-3748

[31] Xiao JQ, Jiang JS, Chien CL. Giant magnetoresistance in nonmultilayer magnetic systems. Physical Review Letters. 1992;**68**:3749-3752

[32] Hylton TL. Limitations of magnetoresistive sensors based on the giant magnetoresistive effect in granular magnetic composites. Applied Physics Letters. 1993;**62**:2431-2433

[33] Hylton TL, Coffey KR, Parker MA, Howard JK. Giant magnetoresistance at low fields in discontinuous NiFe-Ag multilayer thin films. Science. 1993;**261**:1021-1024

[34] Hylton TL, Coffey KR, Parker MA, Howard JK. Low field giant magnetoresistance in discontinuous magnetic multilayers. Journal of Applied Physics. 1994;**75**:7058-7060

[35] Moodera JS, Kinder LR, Wong TM, Meservey R. Large magnetoresistance at room temperature in ferromagnetic thin film tunnel junctions. Physical Review Letters. 1995;**74**:3273-3276

[36] Miyazaki T, Tezuka N. Giant magnetic tunneling effect in $Fe/Al_2O_3/Fe$ junction. Journal of Magnetism & Magnetic Materials. 1995;**139**:L231-L234

[37] Parkin SSP, Kaiser C, Panchula A, Rice PM, Hughes B, Samant M, Yang SH. Giant tunnelling magnetoresistance at room temperature with MgO (100) tunnel barriers. Nature Materials. 2004;**3**:862-867

[38] Wang KL, Kou X, Upadhyaya P, Fan Y, Shao Q, Yu G, Amiri PK. Electric-field control of spin-orbit interaction for low-power spintronics. Proceedings of the IEEE. 2016;**104**:1974-2008

[39] Heimbrodt W, Klar PJ, Ye S, Lampalzer M, Michel C, Baranovskii SD, Thomas P, Stolz W. Magnetic interactions in granular paramagnetic–ferromagnetic GaAs: Mn/MnAs hybrids. Journal of Superconductivity and Novel Magnetism. 2005;**18**:315-320

[40] Lin JH, Wu YQ, Tang S, Fan YL, Yang XJ, Jiang ZM, Zou J. Composition and strain measurements of Ge(Si)/Si(001) islands by HRTEM. Journal of Nanoscience and Nanotechnology. 2009;**9**:2753-2757

[41] Dedkov YS, Holder M, Mayer G, Fonin M, Preobrajenski AB. Spin-resolved photoemission of a ferromagnetic MnSGe3(0001) epilayer on Ge(111). Journal of Applied Physics. 2009;**105**:073909-073912

[42] Petit M, Michez L, Dutoit CE, Bertaina S, Dolocan VO, Heresanu V, Stoffel M, Thanh VL. Very low-temperature epitaxial growth of Mn5Ge3 and Mn5Ge3C0.2 films on Ge(111) using molecular beam epitaxy. Thin Solid Films. 2015;**589**:427-432

[43] Jamet M, Barski A, Devillers T, Poydenot V, Dujardin R, Bayle-Guillemaud P, Rothman J, Bellet-Amalric E, Marty A, Cibert J, Mattana R, Tatarenko S. High-Curie-temperature ferromagnetism in self-organized $Ge_{1-x}Mn_x$ nanocolumns. Nature Materials. 2006;**5**:653-659

[44] Devillers T, Jamet M, Barski A, Poydenot V, Bayle-Guillemaud P, Bellet-Amalric E, Cherifi S, Cibert J. Structure and magnetism of self-organized $Ge_{1-x}Mn_x$ nanocolumns on Ge(001). Physical Review B. 2007;**76**:205306-205315

[45] Park YD, Wilson A, Hanbicki AT, Mattson JE, Ambrose T, Spanos G, Jonker BT. Magnetoresistance of Mn:Ge ferromagnetic nanoclusters in a diluted magnetic semiconductor matrix. Applied Physics Letters. 2001;**78**:2739-2741

[46] Nie T, Kou X, Tang J, Fan Y, Lang M, Chang L-T, Chu C-P, He L, Lee S-W, Xiu F. Superlattice of Fe_xGe_{1-x} nanodots and nanolayers for spintronics application. Nanotechnology. 2014;**25**:505702-505708

[47] Wang Y, Zou J, Zhao Z, Han X, Zhou X, Wang KL. Mn behavior in Ge[sub 0.96]Mn[sub 0.04] magnetic thin films grown on Si. Journal of Applied Physics. 2008;**103**:066104-066103

[48] Wang Y, Xiu F, Wang Y, Xu H, Li D, Kou X, Wang KL, Jacob AP, Zou J. Effect of Mn concentration and growth temperature on nanostructures and magnetic properties of $Ge_{1-x}Mn_x$ grown on Si. Journal of Crystal Growth. 2010;**312**:3034-3039

[49] Xiu F, Wang Y, Kou X, Upadhyaya P, Zhou Y, Zou J, Wang KL. Synthesis of High-Curie-Temperature Fe0.02Ge0.98 quantum dots. Journal of The American Chemical Society. 2010;**132**:11425-11427

[50] Xiu F, Wang Y, Kim J, Hong A, Tang J, Jacob AP, Zou J, Wang KL. Electric-field-controlled ferromagnetism in high-Curie-temperature Mn0.05Ge0.95 quantum dots. Nature Materials. 2010;**9**:337-344

[51] Wang Y, Xiu F, Wang Y, Zou J, Beyermann WP, Zhou Y, Wang KL. Coherent magnetic semiconductor nanodot arrays. Nanoscale Research Letters. 2011;**6**:134-141

[52] Chen YW, Pan BY, Nie TX, Chen PX, Lu F, Jiang ZM, Zhong ZY. Enhanced photoluminescence due to lateral ordering of GeSi quantum dots on patterned Si(001) substrates. Nanotechnology. 2010;**21**:175701-175704

[53] Park Y, Hanbicki A, Erwin S, Hellberg C, Sullivan J, Mattson J, Ambrose T, Wilson A, Spanos G, Jonker B. A group-IV ferromagnetic semiconductor: Mn_xGe_{1-x}. Science. 2002;**295**:651-654

[54] Nie T, Kou X, Tang J, Fan Y, Lee S, He Q, Chang L-T, Murata K, Gen Y, Wang KL. Nanoengineering of an Si/MnGe quantum dot superlattice for high Curie-temperature ferromagnetism. Nanoscale. 2017;**9**:3086-3094

[55] Faxian X, Yong W, Jiyoung K, Hong A, Jianshi T, Jacob AP, Jin Z, Wang KL. Electric-field-controlled ferromagnetism in high-Curie-temperature Mn 0.05Ge 0.95 quantum dots. Nature Materials. 2010;**9**:337-344

[56] Chang L-T, Wang C-Y, Tang J, Nie T, Jiang W, Chu C-P, Arafin S, He L, Afsal M, Chen L-J, Wang K L. Electric-field control of ferromagnetism in Mn-doped ZnO nanowires. Nano Letters. 2014;**14**:1823-1829

[57] Ohno H, Chiba D, Matsukura F, Omiya T, Abe E, Dietl T, Ohno Y, Ohtani K. Electric-field control of ferromagnetism. Nature. 2000;**408**:944-946

[58] Iye Y, Oiwa A, Endo A, Katsumoto S, Matsukura F, Shen A, Ohno H, Munekata H. Metal–insulator transition and magnetotransport in III–V compound diluted magnetic semiconductors. Materials Science and Engineering: B. 1999;**63**:88-95

[59] Li AP, Zeng C, van Benthem K, Chisholm MF, Shen J, Nageswara Rao SVS, Dixit SK, Feldman LC, Petukhov AG, Foygel M, Weitering HH. Dopant segregation and giant magnetoresistance in manganese-doped germanium. Physical Review B. 2007;**75**:201201-201204

[60] Li AP, Wendelken JF, Shen J, Feldman LC, Thompson JR, Weitering HH. Magnetism in Mn_xGe_{1-x} semiconductors mediated by impurity band carriers. Physical Review B. 2005;**72**:195205-195213

[61] Haas C. Spin-disorder scattering and magnetoresistance of magnetic semiconductors. Physical Review. 1968;**168**:531-538

[62] Huang Y, Kang W, Zhang X, Zhou Y, Zhao W. Magnetic skyrmion-based synaptic devices. Nanotechnology. 2017;**28**:08LT02-08LT08

Chapter 5

Double Perovskite Sr_2FeMoO_6: A Potential Candidate for Room Temperature Magnetoresistance Device Applications

Nitu Kumar, Geetika Khurana, Ram S. Katiyar,
Anurag Gaur and R. K. Kotnala

Additional information is available at the end of the chapter

http://dx.doi.org/10.5772/intechopen.70193

Abstract

The family of double perovskites first received attention in the 1960s, but the discovery of low field magnetoresistnace (LFMR) and half metallicity of the Sr_2FeMoO_6 (SFMO) compound was made by Kobayashi et al. in 1998. A fully spin polarized half-metal SFMO (T_c > 400) with excellent magnetoresistance response relatively at small applied fields and high temperatures makes SFMO an ideal candidate for room temperature spintronics applications. Primarily, most of the research work on double perovskites SFMO has been focused on bulk ceramic samples and aimed to understand their structural, magnetic, and magneto-transport properties, along with correlation among them. A material such as SFMO that exhibits a large decrease in resistivity and magnetically order well above room temperature is necessary for the advancement of spintronic devices. If the bulk properties observed could be reproduced in thin films, industrially produced SFMO-based spintronic devices could become a reality. Therefore, the purpose of this chapter is to present the detailed background and descriptions of the double perovskite Sr_2FeMoO_6 (SFMO) thin films and heterostructures with main emphasis to improve or achieve room temperature magnetoresistance properties especially for room temperature magnetoresistive device applications.

Keywords: double perovskite Sr_2FeMoO_6, thin films, magnetic tunnel junctions, tunneling magnetoresistance

1. Introduction

Spintronics is a field of research exploiting the influence of electron spin on the electrical conduction. Spintronics materials have the unique possibilities for use in new functional microelectronic devices and adequate potential to become the ideal memory media for computing and it is a step in the direction of quantum computing due to its advancement in nonvolatility and magnetic random access memory (MRAM). This technology could also be used to create electronic devices, which are smaller, faster and consume less power. Spintronics is one of the emerging technology, which has extended the Moore's law and industry is trying to put more than Moore. Any technology can replace the current world of electronics if it reduces any one of the very large scale integration (VLSI) cost functions like area, power consumption, speed, etc. Fortunately, spintronics can reduce heat dissipation significantly. In charge-based device to switch from logic "0" to logic "1," the magnitude of the charge must be changed in the active region of the device due to which current flows from source to drain. It is not possible with charge-based electronics to reduce the power (or heat) dissipation, since charge is a scalar quantity and the presence or absence of charge gives logic "1" or logic "0." Therefore, to meet the objective scientific community is developing the novel kind of materials that relies on magnetism instead of the flow of current through electron. The first widely acknowledged breakthrough in spintronics was the exploitation of giant magnetoresistance (GMR), a technology, which is now employed in the read heads of most hard drives. The discovery of giant magnetoresistance (GMR) has been cited as the first demonstration of a spintronics application and has been awarded the Nobel Prize in Physics in 2007 [1, 2]. GMR is a quantum mechanical effect observed in thin film heterostructures formed by alternating ferromagnetic and nonmagnetic (NM) layers. When a magnetic field (H) is applied, the thickness of NM layer is chosen such that there is a change in the direction of magnetization in another layer, which reflects a huge change in resistance. That is why the effect is called GMR, a large change in electrical resistance in presence of a magnetic field [2–4]. Baibich et al. represented the GMR of Fe/Cr magnetic superlattices by varying the magnetic field, thickness of NM (Cr) layer and by varying the number of superlattice structure [2]. They reported a large change in resistance or resistivity by applying the small magnetic field, which has a wide application in designing MRAM memories, magnetic read heads, MEMS device, etc. After that, a lot of experiments have been carried out using polarized neutron reflectometry (PNR) tool which clearly illustrates Fe/Cr superlattices that led to GMR effect [5] but PNR has some limitations also [6]. First magnetic sensor using GMR was released in 1994 [7], later, IBM produced the first GMR-read heads for reading data stored in magnetic hard disks [8, 9]. The first GMR-based RAM chips were produced by Honeywell in 1997. Today, GMR-based read heads are frequently used in laptops/computers, iPods, CD/DVD player, and other portable devices. In the twenty-first century, tunneling magnetoresistance (TMR)-based read heads began to displace GMR-based read heads. MRAM chips based on TMR devices are now marketed by several companies, such as Freescale, SanDisk, etc. Current efforts in designing and manufacturing spintronics devices is to optimize the existing GMR-based technology by either developing new materials with larger spin polarization of electrons or making improvements or variations in the existing devices that allow for better spin filtering and try to find new ways to generate and utilize the spin-polarized currents. Till date, magnetic multilayers using giant magnetoresistance (GMR) and tunneling magnetoresistance (TMR) have dominated the data storage industry for many years using simple ferromagnetic metals such as iron and chromium. The investigation of manganite-based magnetic tunnel junctions (MTJs) has deepened our understanding of spin-polarized tunneling and the interface properties of these complex oxides. However, the initial hopes of using them for room-temperature spintronics applications have not been satisfied. Several attempts to replace current microelectronic devices with nanoscale devices have led to a search for new materials with multifunctional properties (multitasking materials that can be manipulated by independent sources). Therefore, many potential half-metallic materials have been predicted and investigated [10–19], although to date roadblocks have occurred in each case. Half-metallic materials are of great interest due to their wide variety of physical properties, including ferromagnetism, ferroelectricity, superconductivity, and many more. In the last few decades, there has been a spectacular enhancement in research activities related to doped manganites, sparked by the discovery of colossal magnetoresistance (CMR) in lanthanum-doped manganites such as $La_{1-x}Ca_xMnO_3$ [20]. The large CMR effect of the order of 10^3 percent is observed at large magnetic fields of several Tesla at low temperature. One of the first working devices using CMR materials was constructed by Sun et al. in 1996 [21]. That consists of two layers of ferromagnetic $La_{0.67}Ca_{0.33}MnO_3$ compound, separated by a thin $SrTiO_3$ spacer layer, which showed a resistance decreased by a factor of 2 in a field of less than 20 mT. However, the main disadvantage of this device was the grain-boundary assisted magnetoresistance properties or in ferromagnetic tunneling junctions is that the large magnetic field sensitivities are only achieved at low temperatures. Furthermore, the CMR effect vanishes far below room temperature due to their low Curie temperatures [21, 22], which make their integration to be difficult for room temperature spintronics applications. The CMR devices exploiting some of the transport properties of manganites close to room temperature have however been proposed, such as contactless potentiometers [23] or bolometers [24, 25], but none of these are strictly speaking spintronics devices. The remarkable magnetoresistive (MR) properties at low-temperature in half-metallic manganites soon motivated the search for new half-metals with higher Curie temperatures [26, 27]. Several high Curie temperature compounds have been predicted to be half-metallic in the 1980s, like semi-Heusler alloys (NiMnSb) [10, 28–31], full-Heusler alloys [32, 33], zinc-blende structure materials [34–36], magnetic oxides (e.g., rutile CrO_2 [13, 37–40] and spinel Fe_3O_4 [41, 42]). However, the first spin-polarization measurements of the Fe_3O_4 and Heusler alloys complex structures were disappointing [43]. CrO_2 has both good conductivity and high T_c, but is an unstable metastable phase that makes incorporation into devices very difficult [12]. Therefore, the half-metallic compounds with high spin polarization can dramatically enhance device performance and are required for a new generation of spintronic devices.

Much effort for the discovery of new high-T_c half-metals focused on perovskites, for which a great experience had been accumulated through the study of manganite films and heterostructures. The family of double perovskites first received attention in the 1960s [44–47], but the discovery of low-field magnetoresistnace (LFMR) and half-metallicity of the Sr_2FeMoO_6 (SFMO) compound were investigated by Kobayashi et al. [48]. A fully spin polarized half-metal, SFMO exhibits a high Curie temperature (>400 K) and excellent magnetoresistance response at relatively small applied fields and at high temperatures compared to manganites, making it an ideal candidate for room temperature spintronics applications.

2. Study of double perovskite Sr_2FeMoO_6

Primarily, most of the research works on double perovskites Sr_2FeMoO_6 (SFMO) has been focused on bulk ceramic samples and aimed to understand their structural, magnetic, and magnetotransport properties, along with correlation among them. The magnetoresistive (MR) properties in double perovskite SFMO generally arises from spin-dependent scattering at the grain boundaries. The underlying conduction mechanism is electron tunneling across insulating grain boundaries. Such magnetoresistance is reported to exist in polycrystalline samples in a magnetic field (~0.1 T), which are considerably lower fields than those utilized for manganite-based devices. The large MR, relatively at smaller external magnetic fields and at room temperature, is required in double perovskites compounds from the applications point of view. A material such as SFMO that exhibits a large decrease in resistivity and magnetically order well above room temperature is necessary for the advancement of spintronic devices.

In our studies, we had synthesized the polycrystalline Sr_2FeMoO_6 sample by conventional solid-state reaction method (shown in **Table 1**). In brief, the stoichiometric amounts of high purity oxides and carbonates, such as $SrCO_3$, Fe_2O_3, and MoO_3, were mixed thoroughly as per above-mentioned formula and calcined at 900°C in Argon (Ar) for 10 h. The calcined powder were reground and pressed into thin pellets of uniform size and sintered at 1200°C for 10 h in a gas flow of 5% H_2 and 95% Ar. The details of the have been provided in our earlier reports [49–52].

Figure 1 shows the rietveld fitted X-ray diffraction patterns of the polycrystalline Sr_2FeMoO_6 sample, which confirm the phase purity of the samples without any observable impurity phases. All the observed peaks of double perovskite phase are clearly visible with significant presence of ordering (103, 211) peaks.

Figure 2 shows the scanning electron micrographs with elemental analysis done at two regions: one at the grain and another at the grain boundary, represented by A and B. The Fe/Mo content ratio calculated through EDS results are 94% at the grain and 93% at grain boundary for SFMO sample, which show the good correlation.

Magnetoresistance versus applied magnetic field plots for Sr_2FeMoO_6 samples are shown in **Figure 3** at 300 and 77 K, respectively. The MR values were calculated by using formula, $MR = 100 \times [\rho(H) - \rho(0)]/\rho(0)$, where $\rho(0)$ and $\rho(H)$ are the resistivity of sample without and with magnetic field, respectively. The pristine Sr_2FeMoO_6 sample shows nearly 11 and 22% magnetoresistance values at 300 and 80 K, respectively, at an applied field 0.72 T. The observed 11% magnetoresistance value at room temperature at 0.72 T magnetic fields confirms the good quality of samples. High LFMR values observed in the pristine SFMO sample may be due to the optimization of intergranular (grain boundaries) and intragranular (domain walls and antisite ordering defects) barrier conditions. However, the role of intergranular barriers has the main contribution in magnetoresistance properties for double perovskite Sr_2FeMoO_6 compound [52–54].

http://dx.doi.org/10.5772/intechopen.70193

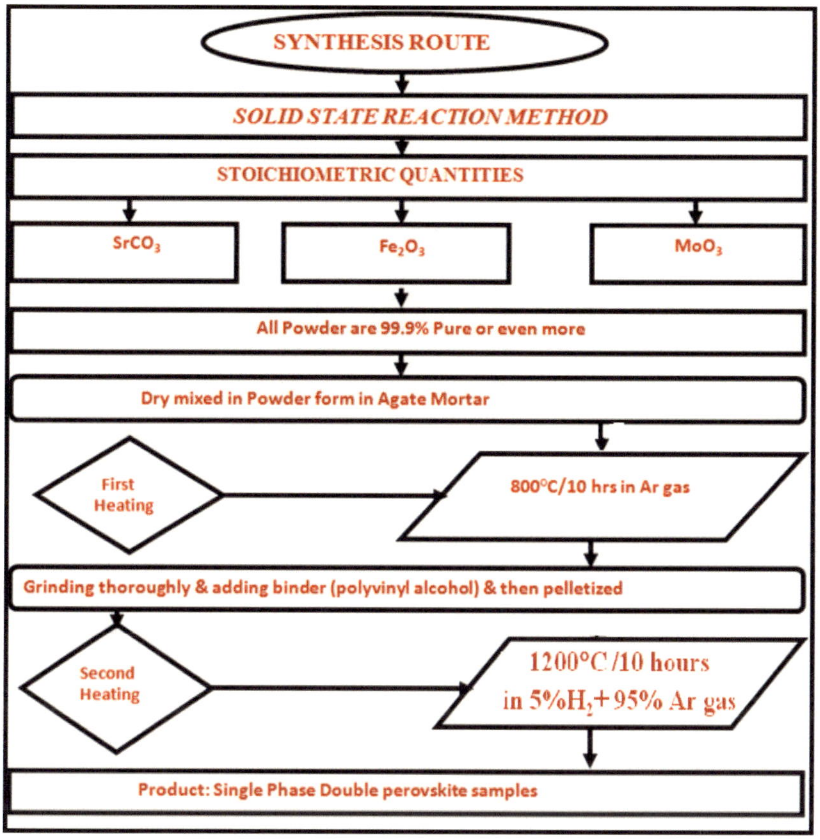

Table 1. Flow chart for the synthesis of Sr_2FeMoO_6 compound.

The observation of high LFMR in such double perovskite system is due to the optimization of the high spin polarization of the carriers and grain boundary adjustment. Recent studies propose a new type of MR, where the spin polarization of grain boundaries is more crucial than the

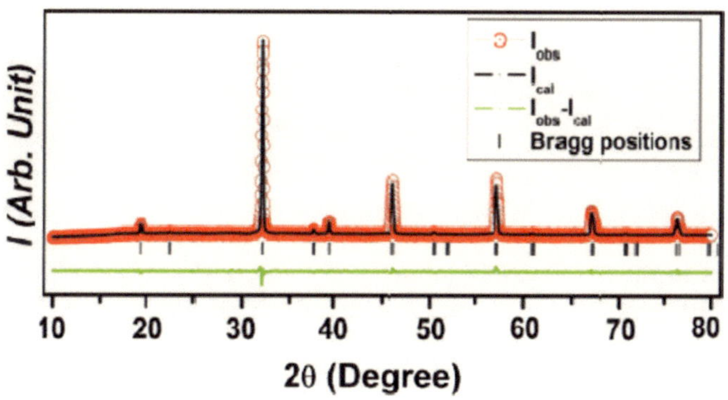

Figure 1. Rietveld fitted X-ray diffraction pattern of the samples Sr_2FeMoO_6.

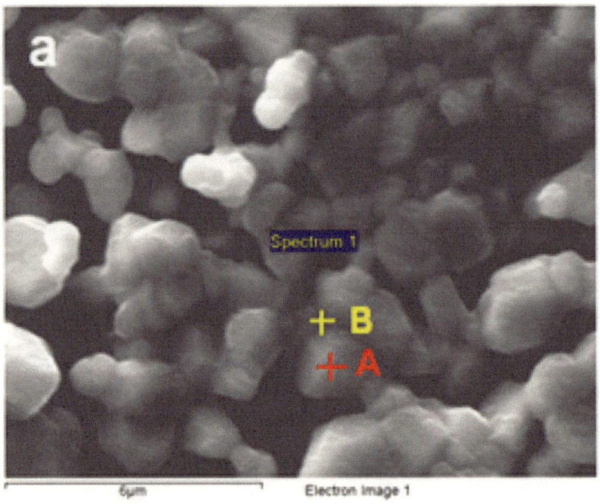

Figure 2. Scanning electron micrographs of Sr_2FeMoO_6 sample. A and B represents the regions where EDS pattern has been recorded at the grain and grain boundaries.

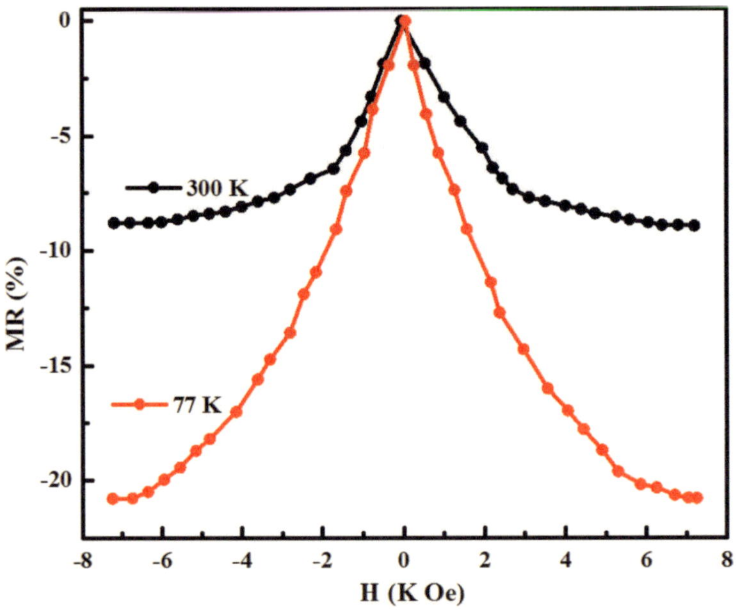

Figure 3. MR plots for the Sr_2FeMoO_6 at 300 and 77 K.

bulk polarization of samples [53, 55, 56]. Furthermore, the grain boundaries are magnetically hard compared to bulk in such double perovskite, which provides a different magnetic switching field for magnetoresistance and magnetization. The different behavior of the curves for normalized $(M/Ms)^2$ versus magnetic field (H) and normalized MR versus H can be observed as one of the indication of such magnetically hard nature of grain boundaries. In our pristine sample such MR does not follow $(M/Ms)^2$, rather $(M/Ms)^2$ versus H curve saturate sharply compared to MR versus H curve, which suggest a MR mechanism similar to as reported by Sarma et al. [53]. Moreover, we estimated the low field behavior of MR and coercive field from magnetization M (μ_B/f.u.) plot and found that the peak in MR is invariably several times larger than the value of coercive field (H_c) for pristine Sr_2FeMoO_6 samples as described by Sarma et al. [53, 56]. This suggests that intragrain properties are not the key determinants for this MR. A remarkable 11% magnetoresistance was observed at room temperature under the presence of low magnetic field (0.72 T). The achieved low-field magnetoresistance value in the sample may be due to the optimized synthesis conditions to get better intergranular tunneling through grain boundaries.

3. SFMO thin films and magnetic tunnel junctions for room temperature magnetoresistive devices

Ultimately, if the bulk properties observed could be reproduced in thin films; industrially produced SFMO-based spintronic devices could become a reality. The most used and optimized method for the growth of SFMO thin film is pulsed laser deposition (PLD) [57–71], however due to inherent complexity of SFMO, its growth as a thin film, has proven to be an arduous task. A brief look through the vast array of SFMO thin film literature reveals that the growth conditions are still in need of perfection [72–83]. The substrate temperature, vacuum, oxygen partial pressure, and gas atmosphere vary from one reference to the next, making it difficult to ascertain the optimal set of growth conditions for SFMO thin films. In the very first study, Manako et al. [71] presented the effect of oxygen pressure and substrate temperature in order to get epitaxial SFMO on $SrTiO_3$ substrates. The phase diagram reported by them demonstrated only a narrow range of oxygen pressures (10^{-5} to 10^{-6} Torr) and temperature higher than 900°C (which is not easy to access by PLD) lead to single-phase, good quality thin films. In contrast, Santiso et al. [72] also grown the SFMO thin films and studied the effect of growth conditions, but found the formation of secondary impurity phases at high growth temperatures. Their results showed that the growth of SFMO thin films at 950°C in ultra-high vacuum (pressure less than 10^{-8} mbar), metallic iron precipitates can form, whereas in a flow of oxygen (pressure of 10^{-6} mbar), iron oxides can be formed. *In situ* X-ray photoelectron spectroscopy (XPS) analysis confirms the presence of secondary phases including $SrMoO_4$ and $SrFeO_3$ on the samples grown at lower pressures (above 10^{-4} mbar). Borges et al. [75] have also grown the SFMO thin films on $SrTiO_3$ substrate at different temperatures and reported significant increment in saturation magnetization (Ms) (1.4–3.5 μ_B/f.u.) by increasing the growth temperature from 770 to 950°C. Similar growth conditions were also used for films grown on slightly larger lattice mismatched MgO substrates. The saturation magnetizations for films were significantly lower than those grown on $SrTiO_3$. These results indicate a direct correlation between substrate temperature and degree of lattice mismatch. In comparison, Song et al. [83] determined that antisite ordering and saturation magnetization were maximized for films grown at 850°C (Ms = 3 μ_B/f.u.). However, substrate temperatures below 850°C and exceeding 900°C resulted in films with a considerable amount of antisite disorder defects. The influence of substrate temperature upon the degree of ordering has also been demonstrated in several studies. While, Westerburg et al. [70] have been able to achieve the highest saturation magnetization (Ms) close to the ideal theoretical value of Ms = 4 μ_B/f.u. for SFMO thin film using combination of high and low growth temperature in PLD. Recently, Venimadhav et al. [73, 74] also obtained ideal value of saturation magnetization Ms = 4 μ_B/f.u. at high growth temperature of 960°C by using the mixture of reducing and oxidizing (O_2 and Ar) atmospheres. It is apparent that the substrate temperature and atmosphere conditions have a large impact on the ordering of SFMO films and should be coupled together to achieve high-quality SFMO thin films.

Another important factor that can affect crystalline quality of the SFMO films is the choice of substrate. The most admired and used substrate for growing SFMO films is $SrTiO_3$ (100) due to the close lattice matching [71, 77, 81, 82]. However, possible presence of oxygen vacancies [74] and low-level magnetic impurities in the $SrTiO_3$ substrate could lead to unusual electrical behavior, potentially causing difficulties in interpreting the data. Other substrates including MgO, $LaAlO_3$ [74, 80, 82, 84], and $NdGaO_3$ [85] have also been used to obtain epitaxial films of SFMO. There have been several studies probing the effects of lattice mismatch on the ordering and magnetization [78, 86]. Asano et al. [86] reported one of the highest magnetizations attained by depositing SFMO (via sputtering deposition) on a lattice matched buffer layer, $Ba_{0.4}Sr_{0.6}TiO_3$, then on $SrTiO_3$ to minimize the effect of 1% lattice mismatch between $SrTiO_3$ and SFMO. A considerable increase in magnetization of 2.3–3.8 μ_B/f.u. and the reduction of the expanded SFMO out-of-plane lattice parameter were observed with the use of the buffer layer. When experiments with the same buffer layer were performed by Sanchez et al. using PLD, slightly lower saturation magnetization value ~3.2 μ_B/f.u. was obtained [78]. In contrast, Yin et al. [80] found their magnetic and transport properties to be independent of the substrates.

The most important and challenging task is to attain good value of MR in SFMO thin films along with the structural and magnetic properties, which will make it ideal for spintronics and magnetoresistive sensors applications. However, epitaxial thin films of SFMO do not show a large MR effect due to lack of grain boundaries. As a means to obtain large MR effects, there have been attempts by varying the deposition conditions and/or the surface of substrates [63, 64, 66]. Therefore, securing a reliable means to fabricate high-quality SFMO thin films and possessing a large MR would be immensely helpful for practical device applications. Manako et al. [71] have reviewed the growth conditions of SFMO thin films, which could only be obtained in a narrow range of deposition temperature and oxygen partial pressure. They studied the magnetoresistance behavior (~5 and ~20% at 300 and 5 K, respectively) of SFMO thin films grown on $SrTiO_3$ (111) and (001) substrates. Their observation showed the large MR effect in (111) oriented films as compared to (001). This might suggest less scattering of carriers at grain boundaries for the (001) oriented film than for polycrystalline samples, since a perfect crystal is expected to show no MR at the temperatures far below T_c. In addition, larger MR of (111) oriented films than that of (100) oriented films may be due to the presence of antiphase domain boundaries in such a way that the superstructure direction is aligned to the growth direction. At the same time, Asano et al. [76] also deposited the Sr_2FeMoO_{6-y} thin films on STO (001) substrate using two-step growth processes. The growth conditions were found to lead either highly conductive metallic thin films or semiconducting films. The metallic films show a positive magnetoresistance (MR) as high as 35%, while the semiconducting films show a small negative MR of −3% at a temperature of 5 K and a field of 8 T. Yin et al. [80] deposited the SFMO epitaxial thin films on $LaAlO_3$ and $SrTiO_3$ (001) substrates. They defined the Wheatstone bridge arrangement straddling a bicrystal boundary to verify that the spin-dependent electron transfer is through a grain boundary or not and found that an intergranular effect is responsible for the LFMR in polycrystalline thin film samples like in the case of bulk polycrystalline samples. Song et al. [63] have also grown the SFMO thin films on $SrTiO_3$ substrates at optimized conditions. They found that the dominant MR mechanism operating in the SFMO films indeed spin-polarized tunneling between the magnetic grains with different orientations across the grain boundaries and concluded that it was possible to achieve a low-field MR value in the film grown at high temperature (935°C), comparable to that of a bulk sample. Shinde et al. [67] have also reported the best quality SFMO films on single-crystalline and polycrystalline $SrTiO_3$ substrates and found that the best quality films could only be grown at particular substrate temperature and oxygen partial pressure. According to their observation, the epitaxial film showed a very small negative MR (1.5% at magnetic field of 8 T), which is almost linear. However, the polycrystalline SFMO film also showed linear type MR behavior with magnitude around (4% at magnetic field of 8 T), which is larger than that of the epitaxial film. Some other efforts have been made to deposit the SFMO thin films in different gas environments and also postannealing treatment to achieve the LFMR by modifying the grain boundary nature [73, 78, 87]. More recently, Saloaro et al. [85] have grown the SFMO thin films on different substrates and observed the absence of traditional magnetoresistance.

Despite the large amount of research reports available for the advancement of magnetoresistance in SFMO thin films, the growth conditions vary from one reference to the next and the results are still controversial. The growth parameters including various substrates and growth temperature can directly affect the surface quality and properties. Furthermore, the fabrication of SFMO-based MTJs structure was rarely reported [61, 68, 88, 89]. There are few groups, including Bibes et al. [68], Asano et al. [88], and Fix et al. [61, 89], who have fabricated the magnetic tunnel junctions and reported the tunneling magnetoresistance (TMR) at very low temperature (5 K). However, it was noticed that the above studies mentioned the growth of MTJ on single crystalline $SrTiO_3$ substrates. In particular, as it was observed that one of the most important considerations for fabrication of the multilayered structure for spintronic device applications is surface/interface quality of the films. High vacuum conditions and *in situ* fabrication of multilayers is generally preferred in order to minimize the amount of secondary phase on the surface of the films. Another obstacle that can slow down the process of using these materials as a source of spin injector is the high growth temperature and complicated growth process. Furthermore, low deposition temperature and silicon substrate is required for their applications in microelectronics industry. Therefore, by considering all above aspects, firstly, we have fabricated the SFMO thin films on Si substrate by pulsed laser deposition and explored the magnetoresistance (MR) behavior besides the structural and magnetic properties. We have optimized the growth conditions and deposited the SFMO thin films at different temperatures ranging from 500–800°C and observed the magnetotransport behavior of SFMO polycrystalline thin films grown on Strontium titanate (STO) buffered Si (100) substrate. After that, we had fabricated the SFMO/$SrTiO_3$/SFMO MTJ structure on $SrTiO_3$ buffered Si (100) substrate to obtain the room temperature magnetoresistance. We observed the ~7% TMR in SFMO/STO/SFMO MTJ, which can be attributed to spin-dependent electron tunneling across the interfaces. The presented results will open up the future prospects of integration of such polycrystalline SFMO thin films for magnetoresistive applications.

A pulsed laser deposition technique was used to deposit the SFMO thin films on $SrTiO_3$ buffered Si (100) substrates. The experimental details are provided below.

1. To optimize the growth conditions of SFMO thin films, substrate temperatures were varied from 500 to 800°C under the base vacuum pressure higher than (>5 × 10^{-6} Torr).
2. Prior to the deposition of SFMO films, STO buffer layer was grown on Si substrate in the presence of high purity (99.99%) oxygen at an ambient pressure of ~50 mTorr at 500°C and subsequently annealed *in situ* at 800°C in O_2 pressure of ~500 mTorr.
3. After the deposition of STO buffer layer on Si (100) substrate, SFMO thin layers were deposited at fixed temperature under the base vacuum pressure (~ 5 × 10^{-6} Torr) and subsequently annealed *in situ* at the same conditions for 10 min.
4. The temperature of substrate was ramped down at 10°C per minute after deposition to prevent thermal shock and cracking of the film.
5. In this study, we fixed all growth parameters and varied only substrate temperature to optimize the growth conditions.

Figure 4, shows the XRD patterns of SFMO thin films grown on STO buffered Si (100) substrate at four different elevated temperatures ranging from 500 to 800°C at a step of 100°C. The growth of SFMO thin film at low temperature (≤ 600°C) do not produce the stoichiometric

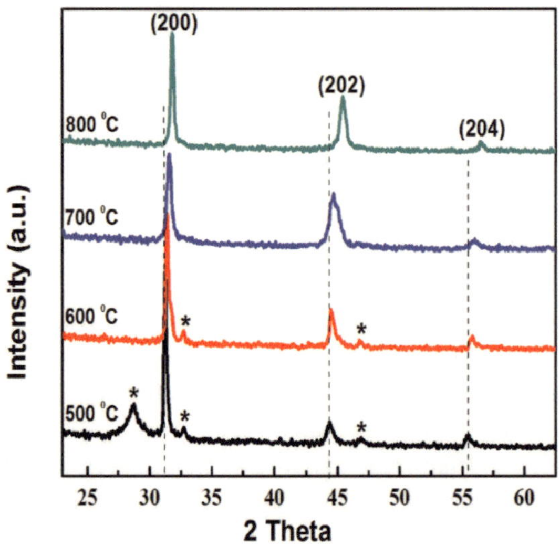

Figure 4. X-ray diffraction pattern of SFMO thin films grown on STO buffered Si (100) substrate at temperatures 500, 600, 700 and 800°C.

SFMO phase and an additional spurious peak of strontium molybdate ($SrMoO_4$) phase (as indicated by * in **Figure 1**) was formed. The secondary impurity phases can be commonly observed at low growth temperatures of SFMO thin films [90, 91]. These impurity phases also develop the antisite disorder in the double perovskite lattice, which further affect the magnetic and magnetotransport properties of SFMO thin films [92, 93]. However, the aforesaid impurity phases were completely disappeared, and a single phase formation of polycrystalline SFMO thin films was observed when the growth temperature was raised to 700°C and above.

The SFMO thin film grown at 800°C shows a series of peaks at 2θ = 31.7, 45.2 and 56.3° corresponding to the (200), (202), and (204) planes, respectively of SFMO phase, which are consistent with the results of Jalili et al. [90]. However, noticeable change in the shifting and sharpening of XRD peaks were observed with increasing growth temperature. This is apparently a consequence of the enhancement in the crystallinity of film with increasing growth temperature. A shifting of peak positions indicates that the thin films are not fully relaxed at low growth temperatures (≤ 600°C). Therefore, we observed that the substrate temperature (T_D) plays crucial role for exact phase formation and improving the crystalline nature of SFMO films.

The magnetoresistance properties of SFMO thin films deposited at different temperatures were measured using four probe resistivity setups at magnetic field up to ±8 T. The magnetoresistnace values are 0.009, 0.017, 0.16, and 0.35% for films deposited at 500, 600, 700, and 800°C, respectively, at room temperature and magnetic field of ±8 T as shown in **Figure 5**.

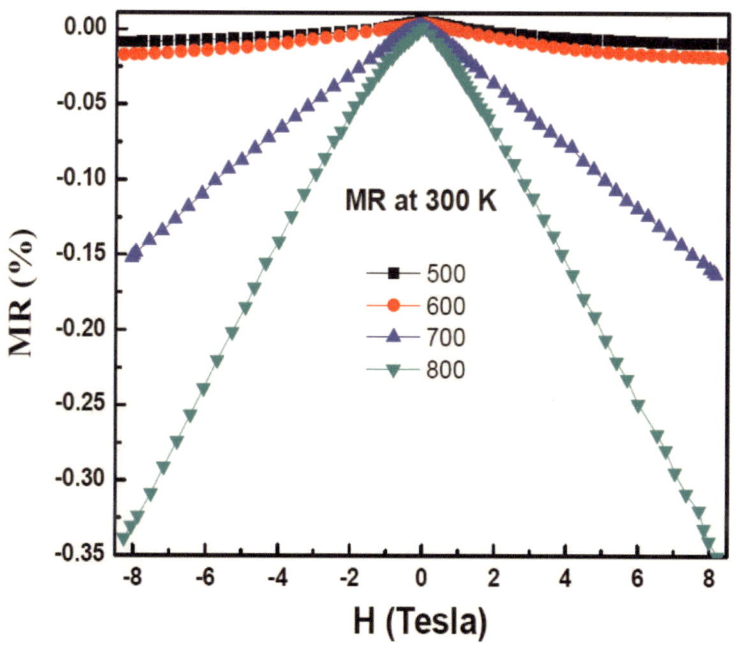

Figure 5. MR behavior of all the SFMO thin films at room temperature.

It is clearly observed that the film grown at low temperature (≤600°C) exhibited the small MR effect and further it was found to be enhanced with growth temperature. This enhanced behavior of MR may be due to the improvement in the crystallinity of the intragranular nature and also by decrement in the antisite disorders effect. Improvement in structural and magnetic properties strongly supports the enhanced magnetoresistance at room temperature. The MR values are 0.35 and 12% at 300 and 5 K for polycrystalline SFMO film grown at 800°C, which shows almost linear type of behavior with magnetic field.

Pulsed laser deposition was used to fabricate SFMO/$SrTiO_3$/SFMO Magnetic tunnel junctions on $SrTiO_3$ buffered Si (100) substrate. The details of the experiments are given below.

1. In the first step, we have optimized the thickness of STO layer by controlling number of laser shots, which was estimated in separate experiments by an empirical relation of thickness and laser pulse counts.
2. The STO buffer layer was grown on Si (100) substrate in the presence of high purity (99.99%) oxygen at an ambient pressure of ~50 mTorr at 500°C and subsequently annealed *in situ* at 800°C in oxygen pressure of ~500 mTorr.
3. The STO buffer layer was characterized by X-ray diffraction technique and found to be crystalline in nature.
4. The SFMO bottom electrode has been deposited at 800°C in the base vacuum pressure higher than ~5×10^{-6} Torr and annealed *in situ* at 800°C for 10 min.
5. The crystal structure and phase purity of SFMO bottom layer was examined by X-ray diffraction and micro Raman microscopy, which shows single-phase formation of SFMO thin film without any impurity phases. The XPS analysis confirms that the Fe:Mo ratio is almost equal (~1:1.1) over the surface of SFMO thin films.
6. Then ~2 nm STO barrier layer was deposited at 800°C with 5×10^{-6} Torr pressure. Again SFMO top layer was deposited at the same conditions as that of bottom SFMO electrode.
7. The shadow mask having lateral dimension 40 μm × 40 μm was used during the fabrication of trilayer SFMO/STO/SFMO structure.

Figure 6(a) shows the schematic of SFMO/STO/SFMO MTJ structure as pattern on STO buffered Si (100) substrate.

The thickness of the SFMO electrodes was kept (≥50 nm) to insure the half-metallic nature of grown SFMO layers [64] and the thickness of STO barrier layer was kept ~2 nm to examine the tunneling effect. The cross-sectional high-resolution field emission scanning electron microscopy (FESEM) image of SFMO/STO/SFMO MTJ structure grown on STO buffered Si (100) substrate as shows in **Figure 6(b)**. The thickness of STO buffer layer is (~10 nm), which is grown on Si substrate. The presence of (~2 nm) STO barrier layer is clearly seen by bright horizontal contrast in the FESEM image, which is perfectly sandwiched between top and bottom SFMO electrodes.

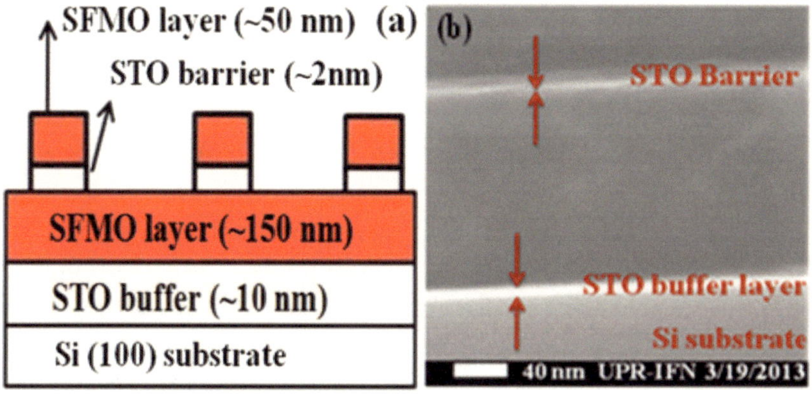

Figure 6. (a) Schematic of the typical SFMO/STO/SFMO MTJ structure. (b) Cross-sectional FESEM image of SFMO/STO/SFMO structure grown on STO buffered Si substrate at 800°C.

Figure 7(a) represents the current voltage (I-V) behavior of SFMO bottom layer and SFMO/STO/SFMO MTJ structure at room temperature. It can be seen from **Figure 7(a)** that I-V behavior of the SFMO bottom layer is almost linear while it shows nonlinear and asymmetric behavior for SFMO/STO/SFMO MTJ at room temperature. This behavior of MTJ is quite different from the SFMO electrode layer and shows the typical characteristic of tunneling assisted transport across a thin insulating barrier [68, 94]. The tunneling conductance in SFMO/STO/SFMO magnetic tunnel junction is further confirmed by fitting the conductance (dI/dV) characteristics as shown in **Figure 7(b)** using Brinkman's formula [95] intended for direct tunneling transport through a rectangular barrier. The thickness of the STO barrier layer was calculated from the fitting of the experimental data and was found to be ~1.5 nm, which is in close agreement with the estimated thickness of the STO barrier layer through FESEM image (~2 nm).

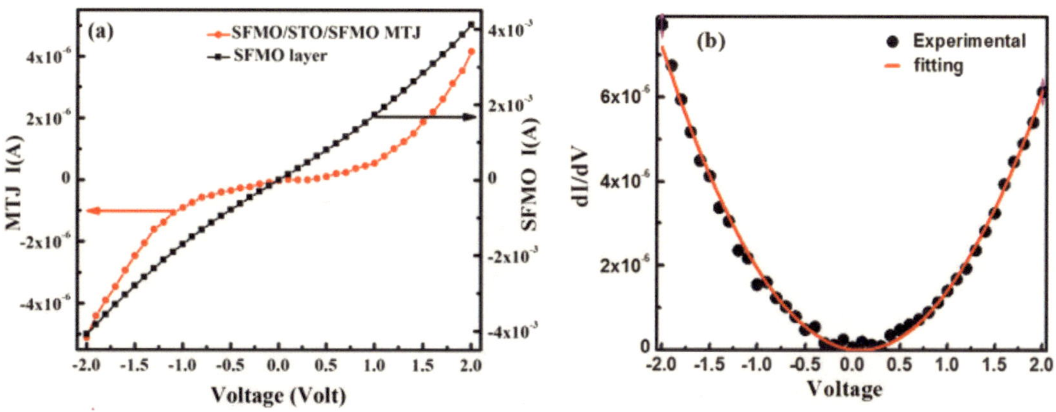

Figure 7. (a) Current-voltage characteristics of SFMO thin films and SFMO/STO/SFMO MTJ at 300 K. (b) Conductance fitting of SFMO/STO/SFMO MTJ.

Figure 8 shows the magnetic field dependence of the resistance for a SFMO/STO/SFMO MTJ with a junction area of 40 × 40 μm², measured at 300 K. This is defined as $(R_{AP} - R_P)/R_P$, where R_P and R_{AP} are the resistances for parallel and antiparallel magnetic configurations of the two electrodes, respectively. When sweeping the field from negative to positive values, the resistance of the junction rises from 2.74 to 2.95 MΩ, yielding a TMR ratio of ~7% at 300 K.

This type of abrupt change in resistance is related to the reversal of two electrodes within the constriction as already observed in Co/I/SFMO [68] and LSMO/STO/LSMO [96] MTJ trilayer junctions fabricated using similar approach. The TMR ratio of MTJ device is related to the spin polarizations P_1 and P_2 of the two ferromagnetic electrodes as TMR = $2P_1P_2/1 - P_1P_2$ using classical Julliere expression [97], where $P = P_1 = P_2$. The observed high value of spin polarization (~18%) at room temperature is attributed to electrons tunneling between SFMO layers through thin (~2 nm) insulating STO barrier. Further to confirm the TMR effect in SFMO/STO/SFMO MTJ, we have independently studied the MR behavior of SFMO film and observed a very small change in MR (~0.35%) at room temperature, which is quite comparable with the reported value for epitaxially grown SFMO films on STO substrate [85, 98]. Based on these results, we propose that the enhanced spin polarization and TMR in SFMO/STO/SFMO MTJ devices is due to the spin-dependant tunneling through ultra-thin insulating and atomically flat STO barrier layer sandwiched between two ferromagnetic SFMO electrodes with sharp interfaces. The sharp change in TMR at the switching field (Hs), where the magnetic moments of the SFMO electrodes realigned from parallel to antiparallel, suggests an entire flip of the magnetic domains against the applied magnetic field. However, the switching field (Hs) at 300 K is less than 30 Oe, which agrees with the coercivity (Hc) of the SFMO film. Hence, the observation of room temperature (~7%) TMR in SFMO/STO/SFMO MTJ can be attributed to spin dependent electron tunneling across the interfaces.

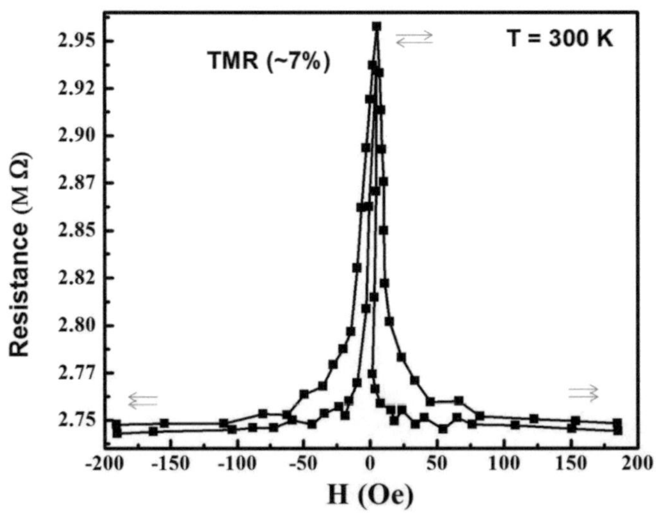

Figure 8. Magnetoresistance behavior of SFMO/STO/SFMO MTJ at 300 K.

http://dx.doi.org/10.5772/intechopen.70193

4. Conclusions

We had synthesized and presented the double perovskite Sr_2FeMoO_6 bulk compound, which shows the remarkable 11% low-field magnetoresistance at room temperature at low magnetic field (0.72 T). This value is one of the high MR value at 300 K, which was only achieved due to the optimized synthesis conditions to achieve better intergranular tunneling through grain boundaries. The brief overview of the synthesis of SFMO thin films has been provided. Furthermore, low deposition temperature and silicon substrate are required for their applications in microelectronics industry; therefore, we have fabricated the SFMO thin films on Si substrate at optimized growth conditions by pulsed laser deposition. The polycrystalline Sr_2FeMoO_6 thin films have been grown on STO buffered Si (100) substrates. We also made an attempt to observe the room temperature magnetoresistance in Sr_2FeMoO_6-based magnetic tunnel junctions (MTJ). The micrometer-sized ($Sr_2FeMoO_6/SrTiO_3/Sr_2FeMoO_6$) devices were grown by pulsed laser deposition. FESEM micrograph analysis revealed the presence of ultrathin (~2 nm) STO barrier layer. Magnetization measurements showed the good ferromagnetic loop behavior with high Curie temperature (T_c) well above 375 K. The current-voltage characteristics of the MTJ devices at room temperature exhibited nonlinear and asymmetric behavior in agreement to the predictions of tunnel conductance. The observed large tunneling magnetoresistance (TMR ~ 7%) at room temperature can be attributed to spin-dependent tunneling across a uniform ultrathin STO tunnel barrier sandwiched between two identical SFMO electrodes in MTJ devices. Finally, we showed that the polycrystalline SFMO thin films and its MTJ structures have enough potential for magnetoresistive and spintronic devices and their possible integration with existing Silicon-based microelectronic devices.

Acknowledgements

Authors acknowledge the financial support from Council of Scientific and Industrial Research Grant at National Physical Laboratory, India and DOD Grant AFOSR-FA9550-16-1-0295 at UPR, Puerto Rico.

Author details

Nitu Kumar[1]*, Geetika Khurana[1], Ram S. Katiyar[1], Anurag Gaur[2] and R. K. Kotnala[3]

*Address all correspondence to: chauhannitu@gmail.com

1 Department of Physics, University of Puerto Rico, San Juan, Puerto Rico, USA

2 Department of Physics, National Institute of Technology, Kurukshetra, India

3 Spintronic and Magnetic Standard Group, CSIR-National Physical Laboratory, New Delhi, India

References

[1] The Nobel Prize in Physics 2007 Albert Fert Unité Mixte de Physique CNRS/THALES, Université Paris-Sud, Orsay, France, and Peter Grünberg Forschungszentrum Jülich, Germany, "for the discovery of Giant Magnetoresistance". http://nobelprize.org/nobel_prizes/physics/laureates/2007/press.html

[2] Baibich MN, Broto JM, Fert A, Van Dau FN, Petroff F, Eitenne P, Creuzet G, Friederich A, Chazelas J. Giant Magnetoresistance of (001)Fe/(001)Cr Magnetic Superlattices. Physical Review Letters. 1988;**61**:2472

[3] Fert A. Nobel Lecture: Origin, development, and future of spintronics. Reviews of Modern Physics. 2008;**80**:1517

[4] Binash G, Grünberg P, Saurenbach F, Zinn W. Enhanced magnetoresistance in layered magnetic structures with antiferromagnetic interlayer exchange. Physical Review B. 1989;**39**:4828

[5] Moyerman1 S, Eckert JC, Borchers JA, Perdue KL, Doucet M, Sparks PD, Carey MJ. Magnetic structure variations during giant magnetoresistance training in spin valves with picoscale antiferromagnetic layers. Journal of Applied Physics. 2006;**99**:08R505

[6] Li M, Cui W, Yu J, Dai Z, Wang Z, Katmis F, Guo W, Moodera J. Magnetic proximity effect and interlayer exchange coupling of ferromagnetic/topological insulator/ferromagnetic trilaye. Physical Review B. 2015;**91**:014427

[7] Daughton JM, Brown J, Chen E, Beech R, Pohm AV, Kude W. Magnetic field sensors using GMR multilayer. IEEE Transactions on Magnetics. 1994;**30**:4608

[8] IBM Archives: IBM 350 disk storage unit. IBM. Retrieved 2011-07-20

[9] Prinz GA. Magnetoelectronics. Science. 1998;**282**:1660

[10] de Groot RA, Mueller FM, van Engen PG, Buschow KHJ. New Class of Materials: Half-Metallic Ferromagnets. Physical Review Letters. 1983;**50**:2024

[11] Pickett WE, Singh DJ. Electronic structure and half-metallic transport in the La1−xCaxMnO3 system. Physical Review B. 1996;**53**:1146

[12] Bibes M, Barthélémy A. Oxide spintronics. IEEE Transactions on Electron Devices. 2007; **54**:1003-1023

[13] Schwarz K. CrO_2 predicted as a half-metallic ferromagnet. Journal of Physics F: Metal Physics. 1986;**16**:L211

[14] Ji Y, Strijkers GJ, Yang FY, Chien CL, Byers JM, Anguelouch A, Gang Xiao and Gupta A. Determination of the Spin Polarization of Half-Metallic CrO_2 by Point Contact Andreev Reflection. Physical Review Letters. 2001;**86**;5585

[15] Brabers VAM. Progress in spinel ferrite research. In: Buschow KHJ, editor. Ferromagnetic Materials. Vol. 8. Amsterdam: Elsevier; 1995. p. 189

[16] Yanase A, Siratori K. Band Structure in the High Temperature Phase of Fe_3O_4. Journal of the Physical Society of Japan. 1984;**53**;312

[17] Serrate D, DeTeresa JM, Ibarra MR. Magnetoelastic coupling in Sr_2(Fe1−xCrx)ReO_6 double perovskites. Journal of Physics: Condensed Matter. 2007;**19**:023201

[18] Krockenberger Y, Mogare K, Reehuis M, Tovar M, Jansen M, Vaitheeswaran G, Kanchana V, Bultmark F, Delin A, Wilhelm F, Rogalev A, Winkler A, Alff L. Sr_2CrOsO_6 : End point of a spin-polarized metal-insulator transition by 5d band filling. Physical Review B. 2007;**75**:020404

[19] Miura Y, Nagao K, Shirai M. Atomic disorder effects on half-metallicity of the full-Heusler alloys Co_2(Cr1−xFex)Al: A first-principles study. Physical Review B. 2004;**69**:144413

[20] Jin S, Tiefel TH, McCormack M, Fastnacht RA, Ramesh R, Chen LH. Thousandfold Change in Resistivity in Magnetoresistive La-Ca-Mn-O Films. Science. 1994;**264**:413

[21] Sun JZ, Gallagher WJ, Duncombe PR, Krusin-Elbaum L, Altman RA, Gupta A, Lu Yu, Gong GQ, Xiao G. Observation of large low-field magnetoresistance in trilayer perpendicular transport devices made using doped manganate perovskites. Applied Physics Letters. 1996;**69**;3266

[22] Yin HQ, Zhou JS, Goodenough JB. Near-room-temperature tunneling magnetoresistance in a trilayer La0.67Sr0.33MnO3/La0.85Sr0.15MnO3/La0.67Sr0.33MnO3 device. Applied Physics Letters. 2000;**77**:714

[23] Balcells LI, Cifre J, Calleja A, Fontcuberta J, Varela M, Benitez F. Room-temperature magnetoresistive sensor based on thick films manganese perovskite. Sensors and Actuators. 2000;**81**:64

[24] Goyal A, Rajeswari M, Shreekala R, Lofland SE, Bhagat SM, Boettcher T, Kwon C, Ramesh R, Venkatesan T. Material characteristics of perovskite manganese oxide thin films for bolometric applications. Applied Physics Letters. 1997;**71**:2535

[25] Yang CH, Koo J, Song C, Koo TY, Lee KB, Jeong YH. Resonant x-ray scattering study on multiferroic BiMnO3. Physical Review B. 2006;**73**:224112

[26] de Teresa JM, Barthelemy A, Fert A, Contour JP, Montaigne F, Seneor P. Role of Metal-Oxide Interface in Determining the Spin Polarization of Magnetic Tunnel Junctions. Science. 1999; **286**:507

[27] Pantel D, Goetze S, Hesse D, Alexe M. Reversible electrical switching of spin polarization in multiferroic tunnel junctions. Nature Materials. 2012;**11**:289

[28] de Groot RA, Mueller FM, van Engen PG, Buschow KHJ. Half-metallic ferromagnets and their magneto-optical properties. Journal of Applied Physics. 1984;**55**:2151

[29] Hanssen KEHM, Mijnarends PE. Positron-annihilation study of the half-metallic ferromagnet NiMnSb: Theory. Physical Review B. 1986;**34**:5009

[30] Galanakis I, Ostanin S, Alouani M, Dreysse H, Wills JM. Ab initio ground state and L2,3 x-ray magnetic circular dichroism of Mn-based Heusler alloys. Physical Review B. 2000;**61**:4093

[31] Fang CM, de Wijs GA, de Groot RA. Spin-polarization in half-metals. Journal of Applied Physics. 2002;**91**;8340

[32] Galanakis I. Surface Properties of the Half- and Full-Heusler Alloys. Journal of Physics: Condensed Matter. 2002;**14**:6329

[33] Galanakis I, Dederichs PH, Papanikolaou N. Slater-Pauling behavior and origin of the half-metallicity of the full-Heusler alloys. Physical Review B. 2002;**66**:174429

[34] Liu BG. Robust half-metallic ferromagnetism in zinc-blende CrSb. Physical Review B. 2003;**67**;172411

[35] Xie WH, Xu YQ, Liu BG, Pettifor DG. Half-Metallic Ferromagnetism and Structural Stability of Zincblende Phases of the Transition-Metal Chalcogenides. Physical Review Letters. 2003;**91**:037204

[36] Galanakis I. Surface half-metallicity of CrAs in the zinc-blende structure. Physical Review B. 2002;**66**:012406

[37] van Lueken H, de Groot RA. Electronic structure of the chromium dioxide (001) surface. Physical Review B. 1995;**51**:7176

[38] Hwang HY, Cheong SW. Enhanced Intergrain Tunneling Magnetoresistance in Half-Metallic CrO_2 Films. Science. 1997;**278**:1607

[39] Korotin A, Anisimov VI, Khomskii DI, Sawatzky GA. CrO_2: A Self-Doped Double Exchange Ferromagnet. Physical Review Letters. 1998;**80**:4305

[40] Tripathy D, Adeyeye AO. Electronic properties of field aligned CrO_2 powders. Physica B. 2005;**368**:131

[41] de Groot RA, Buschow KH. Recent developments in half-metallic magnetism. Journal of Magnetism and Magnetic Materials. 1986;**54**:1377

[42] Pénicaud M, Silberchoit B, Sommers CB, Kübler J. Calculated electronic band structure and magnetic moments of ferrites. Journal of Magnetism and Magnetic Materials. 1992; **103**:212

[43] Bona GL, Meier F, Taborelli M, Bucher E, Schmidt PH. Spin polarized photoemission from NiMnSb. Solid State Communications. 1985;**56**:391

[44] Patterson F, Moeller C, Ward R. Magnetic Oxides of Molybdenum(V) and Tungsten(V) with the Ordered Perovskite Structure. Inorganic Chemistry. 1963;**2**:196

[45] Galasso F, Douglas FC, Kasper RJ. Relationship Between Magnetic Curie Points and Cell Sizes of Solid Solutions with the Ordered Perovskite Structure. Journal of Chemical Physics. 1966;**44**:1672

[46] Nakagawa T. Magnetic and Electrical Properties of Ordered Perovskite $Sr_2(FeMo)O_6$ and Its Related Compounds. Journal of the Physical Society of Japan. 1968;**24**:806

[47] Sleight AW, Longo J, Ward R. Compounds of Osmium and Rhenium with the Ordered Perovskite Structure. Inorganic Chemistry. 1962;**1**:245

[48] Kobayashi KI, Kimura T, Sawada H, Terakura K, Tokura Y. Room-temperature magnetoresistance in an oxide material with an ordered double-perovskite structure. Nature. 1998;**395**:677

[49] Kumar N, Gaur A, Kotnala RK. Stable Fe deficient Sr2Fe1−δMoO6 (0.0 ⩽ δ ⩽ 0.10) compound. Journal of Alloys and Compounds. 2014;**601**:245

[50] Kumar N, Khurana G, Gaur A, Kotnala RK. Room temperature low field magnetoresistance in Sr2FeMoO6/ZnxFe1-xFe2O4 composites. Journal of Applied Physics. 2013; **114**:053902

[51] Kumar N, Aloysius RP, Gaur A, Kotnala RK. Study of ferromagnetic-metal type Sr2FeMoO6 + xAg (x = 0-10 wt%) composites. Journal of Alloys and Compounds. 2013; **559**:64

[52] Pandey V, Verma V, Bhalla GL, Kotnala RK. Increased low field magnetoresistance in electron doped system Sr0.4Ba1.6-xLaxFeMoO6. Journal of Applied Physics. 2012;**108**: 053912

[53] Sarma DD, Ray S, Tanaka K, Kobayashi M, Fujimori A, Sanyal P, Krishnamurthy HR, Dasgupta C. Intergranular Magnetoresistance in Sr_2FeMoO_6 from a Magnetic Tunnel Barrier Mechanism across Grain Boundaries. Physical Review Letters. 2007;**98**:157205

[54] Huang YH, Yamauchi H, Karppinen M. Competition between intragranular and intergranular tunneling magnetoresistance in polycrystalline Sr_2FeMoO_6. Physical Review B. 2006;**74**:174418

[55] Ray S, Middey S, Jana S, Banerjee A, Sanyal P, Rawat R, Gregoratti L, Sarma DD. Origin of the unconventional magnetoresistance in Sr_2FeMoO6. Europhysics Letters. 2011;**94**:47007

[56] Jana S, Middey S, Ray S. Spin-valve-type magnetoresistance: a generic feature of ferromagnetic double perovskites. Journal of Physics: Condensed Matter. 2010;**22**:346004

[57] Ji WJ, Xu J, Zhang ST, Chen YB. Jian Z, ZB Gua,S H Yao, Y F Chena, Thickness dependent microstructures and properties of Sr2Fe10/9Mo8/9O6 films grown in N_2. Solid State Communications. 2013;**163**:28

[58] Paturi P, Metsanoja M, Huhtinen H. Optimization of deposition temperature and atmosphere for pulsed laser deposited Sr_2FeMoO_6 thin films. Thin Solid Films. 2011;**519**:8047

[59] Jalili H, Heinig NF, Leung KT. Growth evolution of laser-ablated Sr_2FeMoO_6 nanostructured films: Effects of substrate-induced strain on the surface morphology and film quality. Journal of Chemical Physics. 2010;**132**:204701

[60] Suominen T, Raittila J, Paturi P. Pure and fully texturized Sr_2FeMoO_6 thin films prepared by pulsed laser deposition from target made with citrate-gel method. Thin Solid Films. 2009;**517**:5793

[61] Fix T, Barla A, Ulhaq-Bouillet C, Colis S, Kappler JP, Dinia A. Absence of tunnel magnetoresistance in Sr_2FeMoO_6-based magnetic tunnel junctions. Chemical Physics Letters. 2007;**434**:276

[62] Boucher R. Sr_2FeMoO_{6+x}: Magnetic and electrical property dependence on thickness and substrate type. Journal of Magnetism and Magnetic Materials. 2006;**301**:251

[63] Song JH, Park JH, Jeong YH. Achieving large magnetoresistance in Sr_2FeMoO_6 thin films. Journal of Applied Physics. 2005;**97**:046105

[64] Fix T, Stoeffler D, Colis S, Ulhaq C, Versini G, Vola JP, Huber F, Dinia A. Effects of strain relaxation on the electronic properties of epitaxial Sr_2FeMoO_6 grown by pulsed laser deposition on $SrTiO_3$ (001). Journal of Applied Physics. 2005;**98**:023712

[65] Sanchez D, Garcıa-Hernandez M, Auth N, Jakob G. Structural, magnetic, and transport properties of high-quality epitaxial Sr_2FeMoO_6 thin films prepared by pulsed laser deposition. Journal of Applied Physics. 2004;**96**:2736

[66] Kim DY, Kim JS, Park BH, Lee JK, Kim JH, lee JH, Chang J, Kim HJ. $SrFeO_3$ nanoparticles-dispersed $SrMoO_4$ insulating thin films deposited from Sr_2FeMoO_6 target in oxygen atmosphere. Applied Physics Letters. 2004;**84**:5037

[67] Shinde SR, Ogale SB, Greene RL, Venkatesan T, Tsoi K, Cheong SW, Millis AJ. Thin films of double perovskite Sr_2FeMoO_6:Growth, optimization, and study of the physical and magnetotransport properties of films grown on single-crystalline and polycrystalline SrTiO3 substrates. Journal of Applied Physics. 2003;**93**:1605

[68] Bibes M, Bouzehouane K, Besse M, Barthelemy A, Fusil S, Bowen M, Seneor P, Contour, JP, Fert A. Tunnel magnetoresistance in nanojunctions based on Sr_2FeMoO_6. Applied Physics Letters. 2003;**83**:2629

[69] Santiso J, Fraxedas J, Balcells L, Fontcuberta J, Figueras A. In situ characterisation of Sr_2FeMoO_6 films prepared by pulsed laser deposition, J. Phys. IV France, 11 PR11 (2001) Pr11-307

[70] Westerburg W, Reisinger D, Jakob G. Epitaxy and magnetotransport of Sr_2FeMoO_6 thin films. Physical Review B. 2000;**62**:R767

[71] Manako T, Izumi M, Konishi Y, Kobayashi KI, Kawasaki M, Tokura Y. Epitaxial thin films of ordered double perovskite Sr_2FeMoO_6. Applied Physics Letters. 1999;**74**:2215

[72] Santiso J, Figueras A, Fraxedas J. Thin films of Sr_2FeMoO_6 grown by pulsed laser deposition: preparation and characterization. Surface and Interface Analysis. 2002;**33**:676

[73] Venimadhav A, Sher F, Attfield JP, Blamire MG. Oxygen assisted deposition of Sr_2FeMoO_6 thin films on $SrTiO_3$(100). Journal of Magnetism and Magnetic Materials. 2004;**269**:101

[74] Venimadhav A, Vickers M, Blamire M. Ferromagnetic spin ordering in disordered Sr_2FeMoO_6 films. Solid State Communications. 2004;**130**:631

[75] Borges RP, Lhostis S, Bari MA, Versluijs JJ, Lunney JG, Coey JMD, Besse M, Contour JP. Thin films of the double perovskite Sr_2FeMoO_6 deposited by pulsed laser deposition. Thin Solid Films. 2003;**429**:5

[76] Asano H, Ogale SB, Garrison J, Orozco A, Li YH, Li E, Smolyaninova V, Galley C, Downes M, Rajeswari M, Ramesh R, Venkatesan T. Pulsed-laser-deposited epitaxial Sr_2FeMoO_6–ySr_2FeMoO_6–y thin films: Positive and negative magnetoresistance regimes. Applied Physics Letters. 1999;**74**:3696

[77] Besse M, Pailloux F, Barthelemy A, Bouzehouane K, Fert A, Olivier J, Durand O, Wyczisk F, Bisaro R, Contour JP. Characterization methods of epitaxial Sr_2FeMoO_6 thin films. Journal of Crystal Growth. 2002;**241**:448

[78] Sanchez D, Auth N, Jakob G, Martinez JL, Garcia-Hernandez M, et al. Pulsed laser deposition of Sr_2FeMoO_6 thin films. Journal of Magnetism and Magnetic Materials. 2005;**294**:e119

[79] Fix T, Versini G, Loison JL, Colis S, Schmerber G, Pourroy G, Dinia A. Pressure effect on the magnetization of Sr_2FeMoO_6 thin films grown by pulsed laser deposition. Journal of Applied Physics. 2005;**97**:024907

[80] Yin HQ, Zhou JS, Zhou JP, Dass R, McDevitt JT, Goodenough JB. Intra- versus intergranular low-field magnetoresistance of $Sr_2FeMoO_6Sr_2FeMoO_6$ thin films. Applied Physics Letters. 1999;**75**:2812

[81] Trolio AD, Larciprete R, Testa A, Fiorani D, Imperatori P, Turchini S, Zema N. Double perovskite $Sr_2FeMoO_6Sr_2FeMoO_6$ films: Growth, structure, and magnetic behavior. Journal of Applied Physics. 2006;**100**:0139071

[82] Wang S, Pan H, Zhang X, lian G, Xiong G. Double-perovskite Sr_2FeMoO_6 epitaxial films with ordered cation structure grown in mixture gas of hydrogen and argon. Applied Physics Letters. 2006;**88**:121912

[83] Song JH, Park BG, Park JH, Jeong YH. Double-Perovskite Sr_2FeMoO_6 Thin Films Prepared by Using Pulsed Laser Deposition: Growth and Crystal, Electronic and Magnetic Structures. Journal of the Korean Physical Society. 2008;**53**:1084

[84] Jalili H, Heinig NF, Leung KT. X-ray photoemission study of Sr_2FeMoO_6 and $SrMoO_4$ films epitaxially grown on MgO(001): Near-surface chemical-state composition analysis. Physical Review B. 2009;**79**:174427

[85] Saloaro M, Majumdar S, Huhtinen H, Paturi P. Absence of traditional magnetoresistivity mechanisms in Sr_2FeMoO_6 thin films grown on $SrTiO_3$, MgO and $NdGaO_3$ substrates. Journal of Physics: Condensed Matter. 2012;**24**:366003

[86] Asano H, Kohara Y, Matsui M. Coherent Epitaxy and Magnetic Properties of Sr_2FeMoO_6 Thin Films on Ba0.4Sr0.6TiO3-Buffered $SrTiO_3$ Substrates. Japanese Journal of Applied Physics. 2002;**41**:L1081

[87] Metsanoja M, Majumdar S, Huhtinen H, Effect of ex situ post-annealing treatments on Sr_2FeMoO_6 thin films. Journal of Superconductivity and Novel Magnetism. 2012;**25**:829

[88] Asano H, Koduka N, Imaeda K, Sugiyama M, Matsui M. Magnetic and junction properties of half-metallic double-perovskite thin films. IEEE Transactions on Magnetics. 2005;**41**:2811

[89] Fix T, Stoeffler D, Henry Y, Colis S, Dinia A, Dimopoulos T, Bär L, Wecker J. Diode effect in all-oxide Sr_2FeMoO_6-based magnetic tunnel junctions. Journal of Applied Physics. 2006;**99**:08J107

[90] Jalili H, Heinig NF, Leung KT. Formation of nanocrystalline films of $Sr_2FeMoO_6Sr_2FeMoO_6$ on Si(100) by pulsed laser deposition: Observation of preferential oriented growth. Journal of Applied Physics. 2009;**105**:034305

[91] Bianchini L. Synthesizing and characterizing Sr_2FeMoO_6 bulk and thin films. Undergraduate Review. 2008;**4**:65

[92] Kim DY, Kim JS, Park BH, Lee JK, Kim JH, lee JH, Chang J, Kim HJ. I Kim,Y D Park, $SrFeO_3$ nanoparticles-dispersed $SrMoO_4$ insulating thin films deposited from Sr_2FeMoO_6 target in oxygen atmosphere. Applied Physics Letters. 2004;**84**;5037

[93] Hernandez MG, Martinez JL, Martinez-Lope MJ, Casais MT, Alonso JA. Finding Universal Correlations between Cationic Disorder and Low Field Magnetoresistance in FeMo Double Perovskite Series. Physical Review Letters. 2001;**86**:2443

[94] Yin HQ, Zhou JS, Dass R, Zhou JP, McDevitt JT, Goodenough JB. Grain-boundary room-temperature low-field magnetoresistance in Sr_2FeMoO_6 films. Journal of Applied Physics. 2000;**87**:6761

[95] Brinkman WF, Dynes RC, Rowell JM. Tunneling Conductance of Asymmetrical Barriers. Journal of Applied Physics. 1970;**41**:1915

[96] Bowen M, Bibes M, Barthelemy A, Contour JP, Anane A, Lemaıtre Y, Fert A. Nearly total spin polarization in La2/3Sr1/3MnO3La2/3Sr1/3MnO3 from tunneling experiments. Applied Physics Letters. 2003;**82**:233

[97] Julliere M. Tunneling between ferromagnetic films. Physics Letters A. 1975;**54**:225

[98] Muduli PK, Budhani RC, Topwal D, Sarma DD. Spin-polarized electron tunneling in polycrystalline Sr_2FeMoO_6 thin films. Journal of Physics: Conference Series. 2009;**150**:042132

Chapter 6

Giant Magnetoresistance Sensors Based on Ferrite Material and Its Applications

Mitra Djamal and Ramli Ramli

Additional information is available at the end of the chapter

http://dx.doi.org/10.5772/intechopen.70548

Abstract

In recent decades, new magnetic sensors based on giant magnetoresistance (GMR) have been studied and developed intensively. GMR materials have great potential for next-generation magnetic field sensing devices. The GMR material has many attractive features, for example, its electric and magnetic properties can be varied in a very wide range, low power consumption, and small size. Therefore, GMR material has been developed into various applications of sensor based on magnetic field sensings, such as magnetic field sensor, a current sensor, linear and rotary position sensor, data storage, head recording, nonvolatile magnetic random access memory, and biosensor. In this chapter, the recent development of a GMR thin-film–based ferrite material will be reviewed. Furthermore, recent and future trend application of GMR sensor will be discussed.

Keywords: biosensor, ferrite, giant magnetoresistance, GMR sensor, magnetic sensor

1. Introduction

In recent years, many attempts were made to improve the reliability of the sensors and sensor systems and at the same time lower the cost of fabrication. This aims to make the price of the sensor become relatively cheap. The sensors and sensor systems have been developed for various applications such as in motor vehicles, housing (e.g., for security, regulation of air circulation, temperature regulation, setting humidity), delivery of food, or warehouse storage of food (e.g., temperature, humidity, gas concentration).

In general, a sensor is defined as a device that converts physical, chemical, or biological quantities into electrical quantities. The capability of a sensor or sensor system is determined by the strong interaction of the three main constituent components, such as sensor structure, manufacturing technology, and signal processing algorithms. The development of sensor technology is also influenced by the development of these three areas as shown in **Figure 1**.

IC Insights 2017 [1] has reported that all sensor categories such as pressure sensors, acceleration sensors, and magnetic sensors and most actuators have double-digit sales in 2016. The market for sensors and actuators in 2017 is predicted to increase by 7.8% and hit a record high of $12.8 billion. Sales of sensors/actuators estimated over the next 5 years will be driven by the deployment of automated control functions that are integrated with the vehicle (including autonomous driving capability), unmanned aircraft, systems of industrial and robotics, everyday electronics, and measurement unit related to Internet of Things (IoT).

The need for new sensors and electronic interfaces, particularly in portable applications, which show small dimensions and the ability to reduce both the supply voltage and power consumption, is in continuous growth. In particular, multiple sensors and electronic circuits for interfacing developed in an integrated technology can be combined into just one chip, and enabling it to produce "smart sensor."

The phenomenon of giant magnetoresistance (GMR) has been providing cutting-edge sensor technology, especially for affordable and sensitively detect and quantify of micro-particles and nano-magnetic in very weak magnetic fields. In recent decades, sensors based on GMR effect have been researched and developed intensively [3]. The discovery of GMR has opened

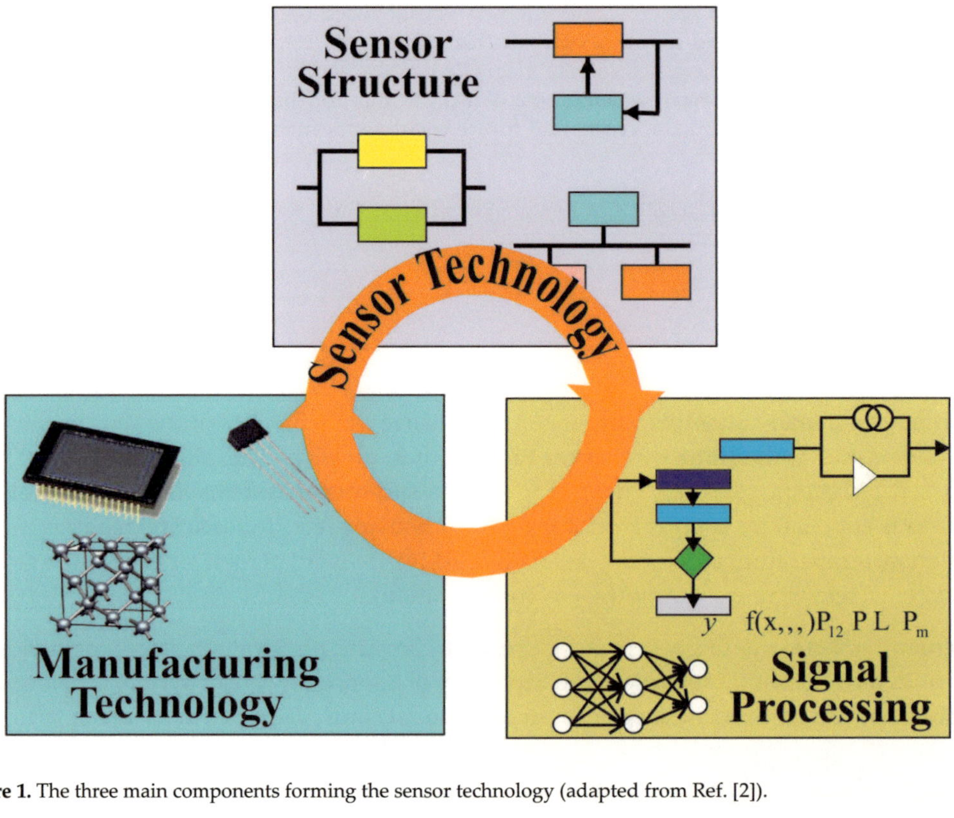

Figure 1. The three main components forming the sensor technology (adapted from Ref. [2]).

opportunities in many fields of applications. GMR material has been developed into various applications of sensor based on magnetic field sensing, such as magnetic field sensor, a current sensor, linear and rotary position sensor, data storage, head recording, and nonvolatile magnetic random access memory. The GMR material has many attractive features, for example, its electric and magnetic properties can be varied in a very wide range, low power consumption, and small size. Meanwhile, ferrite is one of the candidates of magnetic oxide material that could potentially be used as a constituent layer of GMR [4].

This chapter is organized as follows: the magnetic sensor, the GMR sensor based on ferrite material, and the GMR sensor design. Finally, the recent and the future trends of this exciting GMR sensor for various applications are discussed.

2. Magnetic sensor

Research in the magnetic sensor has been carried out by researchers in recent decades. Magnetic sensors have a significant impact over the past five decades in a variety of different fields of technology. Magnetic sensor has great potential to be developed for various applications such as magnetic storage, automotive sensors, navigation systems, nondestructive material testing, security system, structural stability, medical sensors, and military instruments [5].

Based on the measurement range of the magnetic field, approximate sensitivity ranges of different magnetic field sensors are low field (smaller than 0.1 nT), medium field (0.1–1 nT), and high field (above 1 nT), as shown in **Figure 2**.

Figure 2 shows measurement range of the magnetic field of some magnetic sensors. The GMR sensor can detect magnetic fields in the range of 10–10^8 nT and has a die size close to 1 mm.

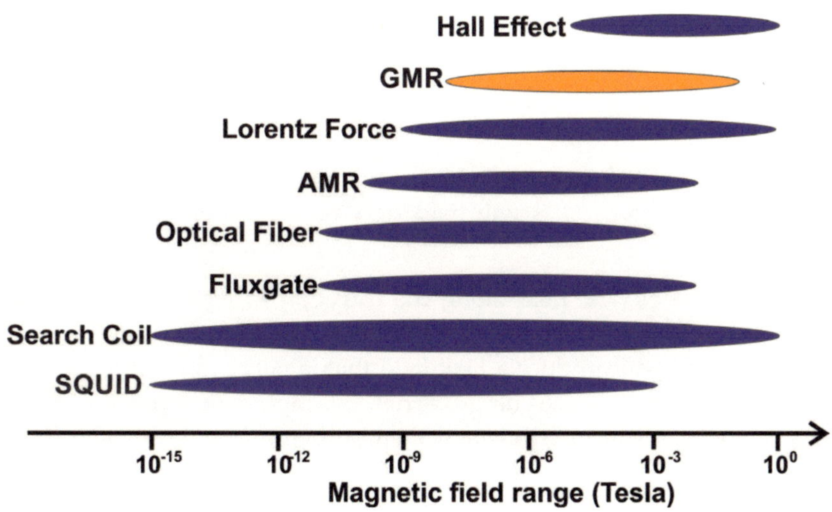

Figure 2. Approximate sensitivity range of magnetic sensor (adopted from Ref. [6]).

The dependence on temperature of the offset and sensitivity of the GMR sensor is higher than the AMR (anisotropic magnetoresistance) sensor. In addition, GMR sensors can operate at temperatures above 225°C [6].

The term "magnetic sensor" is widely used to express a sensor that works on the magnetic principles. Magnetic sensors usually work without contact with the object to be sensed and also reliable. The most important field in the application of magnetic sensors includes security, health care, information technology, geomagnetic exploration, and nanotechnology. Magnetic field sensor technology has been driven by the need to increase the sensitivity, small size, low power, low cost, and compatibility with electronic systems. To achieve these requirements, the magnetic sensors are usually created by micrometer sized or sub-micrometer with a multilayer structure.

The resistance of the material depends on the state of magnetization, called the magnetoresistance effect. Magnetization in the material can be changed by applying an external magnetic field. Therefore, the material which has a magnetoresistance effect can be used as magnetic field sensor.

The basic effect of normal magnetoresistance emerged from the Lorentz force on the electrons due to the presence of a magnetic field applied to it. Normal magnetoresistance occurs in all metals, including nonmagnetic metals as a consequence of Lorentz forces. For example, in metal thick film, cobalt (Co) with a thickness of 100 nm was observed to be positive and varies with B^2 above magnetoresistance saturation, where B is applied in a magnetic field. However, Co thin films with a thickness of 3 nm show negative normal magnetoresistance over saturated magnetoresistance. Normal magnetoresistance emerged from semiclassical arguments via the Lorentz force on electrons that are defined as

$$\vec{F} = m\left(d\vec{v}/dt\right) = e\vec{E} + \left(e\vec{v} \times \vec{B}\right) \tag{1}$$

and the current density

$$\vec{j} = \left(n e^2 \tau/m\right)\vec{E} + \left(e^2 \tau/m\right)\vec{v} \times \vec{B} \tag{2}$$

where $\vec{F}$ = Lorentz force, $\vec{v}$ = electron velocity, $\vec{j}$ = current density, m = electron mass, e = electron charge, $\vec{E}$ = electric field, $\vec{B}$ = magnetic field, n = electron concentration, and τ = relaxation time.

The components of the electric field along the direction of $\vec{j}$ do not change without the presence of the magnetic field $\vec{B}$, so magnetoresistance is normally equal to zero. The presence of a magnetic field will produce a Hall field that is perpendicular to the $\vec{j}$ direction so that the Lorentz force acting on the charge carrier will increase the resistance in a magnetic field. Using the model of free electrons and then normal magnetoresistance is obtained as $(\Delta\rho/\rho) = (R_H/\rho)^2 B^2$, where R_H is the Hall resistance.

Electronic devices in the future will turn on a new field called spintronics. Spintronics is a new field that explores the influence of spin on electronic transport in magnetic nanostructures.

Spintronic births are marked by the discovery of giant magnetoresistance effects (GMR) on magnetic multilayer Fe/Cr more than two decades ago. Accordingly, spintronics is a technology that exploits the quantum property of electrons called spin. Ordinarily, electron spins have both "up" and "down" directions and can be described as clockwise or counterclockwise around their axes. The spin gives magnetic properties on electrons that can be affected by external magnetic fields. More recently, research on spintronic devices developed very rapidly [7–9].

3. The GMR sensor based on ferrite material

The GMR-based magnetic sensor is a sensor that works based on effect of a very large change in the resistance of metal or device when an external magnetic field is applied. The magnetoresistance (MR) ratio value is written in Eq. (3):

$$\mathrm{MR} = \frac{\Delta R}{R} = \frac{R(H) - R(H=0)}{R(H=0)} \tag{3}$$

where $R(H)$ is the resistance when the device is influenced by an external magnetic field, $R(H = 0)$ is the resistance of the device without the applied external magnetic field, and H is the magnetic field intensity.

GMR material could have several structures where each structure will produce different GMR ratios. This structure consists of a sandwich, spin valve (pinned sandwich), multilayer, and granular (**Figure 3**). The sandwich structure is also called pseudo spin valve, which consists of three layers with the arrangement of materials (ferrimagnetic or ferromagnetic)/nonmagnetic/(ferrimagnetic or ferromagnetic). Accordingly, in spin valves, an additional antiferromagnetic (pinning) layer is added to the top or bottom part of the sandwich structure. Meanwhile, a multilayer structure is a structure with repetition of the sandwich layer. Similarly, the granular structure consists of granules of magnetic materials of nanometer scale scattered in nonmagnetic material as the host material.

The benefit of the GMR phenomenon is on the development of nanometer-sized technologies and possibly in the atomic scale of magnetic structures. This very thin structure has physical, chemistry, and biology properties which change dramatically, therefore, superior when compared to the bulk materials. A very thin layer can be made as an epitaxial layer (where the layer has a certain crystal arrangement with excellent monocrystalline quality) by molecular beam epitaxy (MBE) method or polycrystalline film by sputtering method [10].

Meanwhile, measurement of GMR effects has two main geometries, i.e., current In plane (CIP) and current perpendicular to plane (CPP). These two geometric shapes are shown in **Figure 4**.

Until now, researchers have continued to conduct research on GMR thin film, regarding on growth methods, constituent materials, as well as the GMR structure. Particularly for GMR constituents, ferrite is a transition metal oxide and one of the potential candidates used as the constituent material of the GMR thin film [11, 12]. Transition metal oxides are an important functional material for new electronic devices due to their unique properties such as perfect

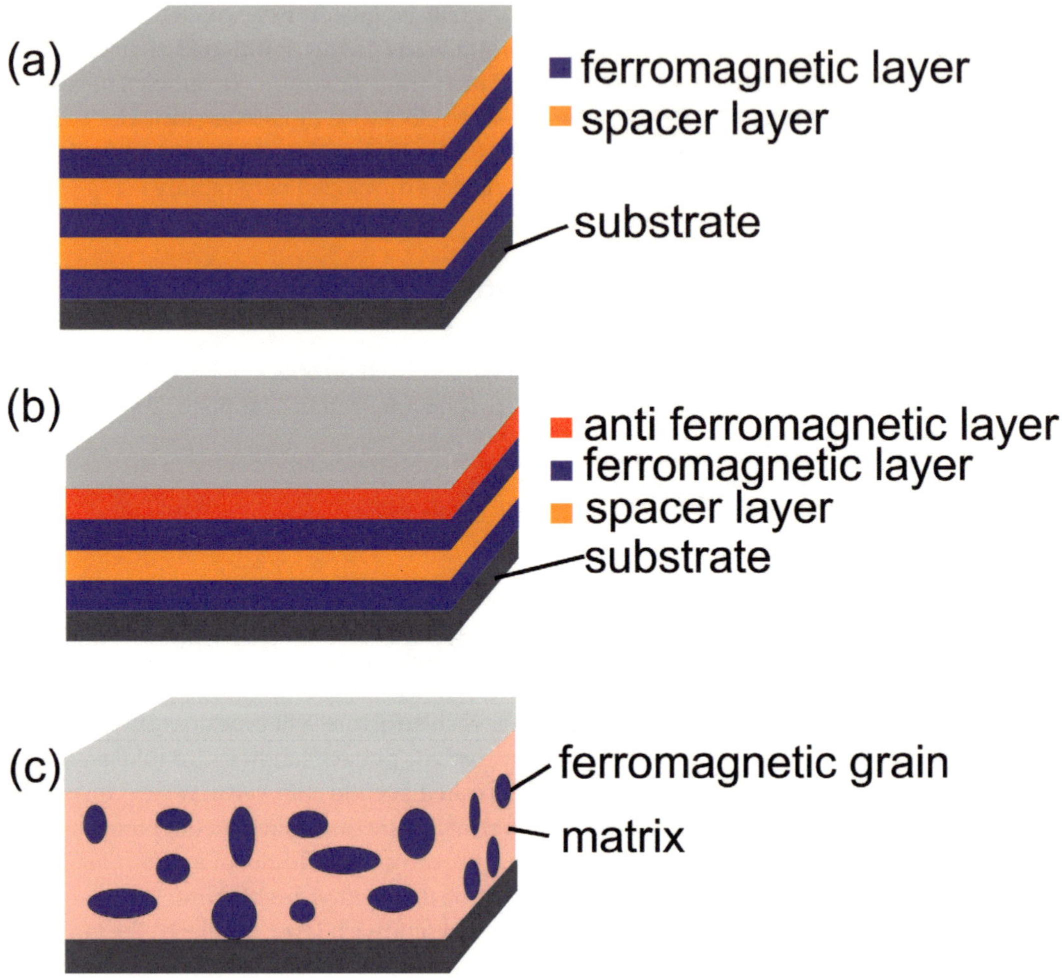

Figure 3. Various types of GMR structures: (a) multilayer, (b) spin valve, and (c) granular films.

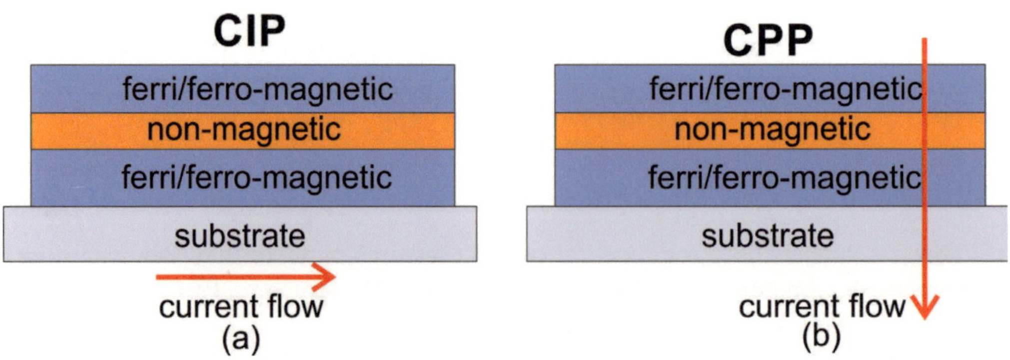

Figure 4. Geometry of GMR measurement (a) CIP and (b) CPP.

spin polarization, large metal insulator transitions, ferroelectric, multiferroic and resistive switching effects [13]. In addition, ferrite has ferrimagnetic properties and Curie temperature above room temperature. For temperatures below the Curie temperature, ferrimagnetic materials exhibit the same behavior as ferromagnetic materials. The behavior is having a spontaneous magnetization at room temperature, consisting of saturated magnetic domains, and shows hysteresis phenomena [14]. Among the ferrite materials that have been used as GMR constituent layers are Fe_3O_4 and $CoFe_2O_4$.

The phenomenon of magnetoresistance has been observed among others the nanowires Fe_3O_4 single crystal [15], Fe_3O_4 thin film [16, 17], Fe_3O_4/GaAs/Fe_3O_4 junction [18], and Fe_3O_4 nanoparticles [19]. Recently, we report the results of studies relating to the synthesis of a novel ternary $CoFe_2O_4$/CuO/$CoFe_2O_4$ thin film as a GMR sensor [20]. The $CoFe_2O_4$/CuO/$CoFe_2O_4$ thin film has been prepared onto silicon substrate via dc magnetron sputtering technique with targets facing each other. The GMR ratio maximum at room temperature obtained has reached 70% for the thickness of each layer of $CoFe_2O_4$ and CuO, which are 62.5 nm and 14.4 nm, respectively. These findings provide the impact on GMR sensor technology based on ferrite material.

4. The GMR sensor design

The performance of the GMR sensor is influenced by various parameters. Among them is the composition of GMR constituent material, layer thickness, and GMR structure. The GMR structure is concerned with applications on applied technology. For example, the condition of the technology affects the performance of the spin-valve structure, so it is quite a concern because this structure is a very promising candidate for sensors and reading heads.

The following features are used to determine a good quality GMR sensor, that is [21], a large magnetoresistance ratio, large sensitivity, narrow hysteresis characterized by a low coercive field (Hc), low anisotropy field (Hk) (Hk influences sensitivity), large exchange bias field (Hex), minor changes of parameters with temperature, and have great reliability and repeatability.

The advantages of GMR sensor have a larger output than AMR sensor or Hall effect sensor and can be operated on the field above the AMR sensor field range. Some advantages of the GMR sensor compared with the AMR sensor or Hall effect sensor are listed in **Table 1**.

Advantages	GMR	AMR	Hall
Physical size	Small	Large	Small
Signal level	Large	Medium	Small
Sensitivity	High	High	Low
Temperature stability	High	Medium	Low
Power consumption	Low	High	Low
Cost	Low	High	Low

Table 1. Benefits of magnetic sensors [22].

The design and development of GMR sensors are based on assessments derived from the various areas to obtain functional devices. In addition, the special design will always be associated with specific applications. The sensor design will have an impact on the device performance. For example, both linear and thermal sensor characteristics affect the characteristics of the sensing structure and the final encapsulation. A more detailed knowledge of the sensor parameters is necessary before embarking on the development of sensors.

As a sensing element, the Wheatstone bridge circuit has been highly recommended in the design of resistive sensors. In this case, the Wheatstone bridge circuit provides a differential output as a function of the resistance variation. Depending on cases deemed or specific requirements, we can use multiple bridge configuration.

The GMR sensor made using the Wheatstone bridge principle aims to reduce the effect of hysteresis and to improve the output linearity. Ordinarily, the GMR sensor structure consists of four resistors. Two resistors are protected from exposure to the magnetic field, and two other active resistors are between two flux concentrators. The sensitivity of the GMR sensor can be changed by changing the length and distance between two flux concentrators.

Figure 5 displays a summary of the possibility of the Wheatstone bridge configuration that has been furnished by calculating the output voltage. As shown in **Figure 5**, a full bridge configuration is the best choice in terms of the signal level and linearity (**Figure 5**, right). Due to the stage dependence on the fabrication process of GMR structure, a half bridge configuration with two active resistors and two shielded resistors is often obtained when using a single-stage deposition. However, full bridge configuration is obtained if using two-step deposition.

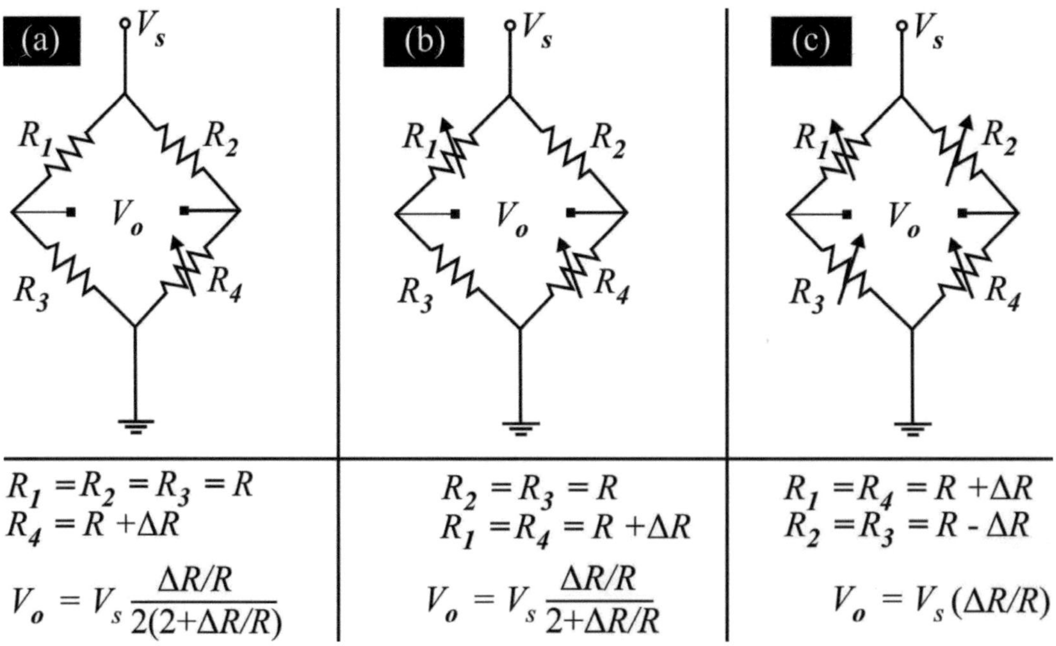

Figure 5. Wheatstone bridge configuration (a) special elements, (b) half bridge, (c) full bridge [23].

In the design of GMR thin-film structures, it is useful to operate at the resistivity determined for the condition $w = L$ (w and L are width and length of GMR thin film, respectively):

$$R_{sq} = \frac{\rho}{t} \tag{4}$$

where R_{sq} = resistance "ohms per square," ρ = resistivity, and t = thickness of the film. One important factor in GMR sensor application is the resolution associated with the ratio of signals to noise. According to Ref. [24], the signal-to-noise ratio (SNR) is estimated to be proportional to the square root of the device area, following Eq. (5):

$$\mathrm{SNR} = c\sqrt{\mathrm{Lw}} \tag{5}$$

where c is a constant.

Nordling et al. [25] have used the Wheatstone bridge to develop integrated GMR sensors. This system consists of four GMRs that are integrated into serpentine, as shown in **Figure 6**. Two GMRs are formed interlocked as folded fingers act as sensing resistors (R_{S1} and R_{S2}), whereas two GMRs are spatially separated as reference resistors (R_{R1} and R_{R2}). The width of each strip is 2 μm, the separation distance is 2 μm, and the total length is more than 11 mm. Accordingly, the edges of the sensing and reference GMRs are separated by 30 μm to each other, to isolate the inbred magnetic field of each resistor, which is a function of current from leads. Hereafter, the GMR sensing elements are coated with a silicon nitride thin film to form an active surface and protect it from being in contact with the sample.

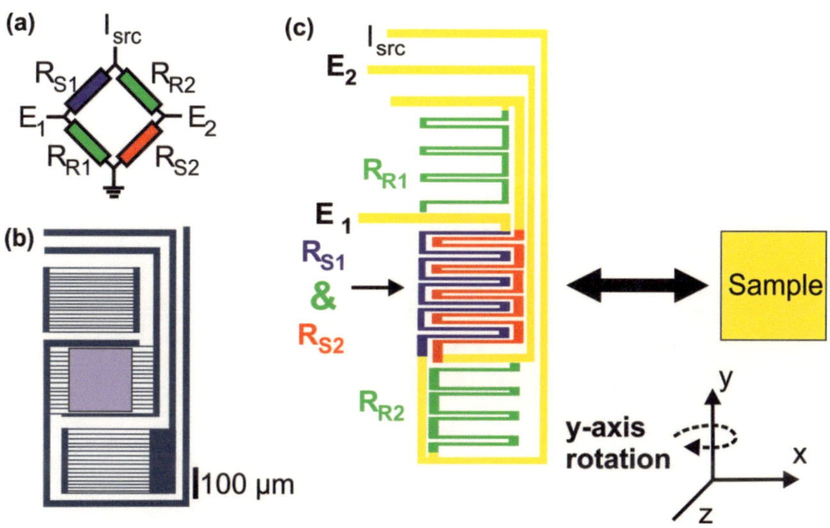

Figure 6. (a) Wheatstone bridge circuit. (b) A micrograph image of GMR sensing showing the reference GMR traces (top and bottom) and the GMR sense pad (center). (c) The schematic of the Wheatstone bridge at (b) in a circuit also shows the direction of the external field and the relative motion of GMR to the sample (adapted from Ref. [25]).

This type of resistive sensor interface of the Wheatstone bridge shows a low sensitivity value and cannot be recovered. This weakness can be corrected by using a differential input OpAmp-based voltage amplifier, to increase the sensitivity as shown in **Figure 7**.

Furthermore, the Wheatstone bridge configuration can be made "automatic" (so the circuit does not require initial calibration) through the development of a topology as shown in **Figure 8**. This circuit using a tunable resistor implemented through a voltage-controlled resistance based on the use of an analog quadrant multiplier. The variations follow those of the resistive sensor and a suitable feedback loop.

More specifically, the circuit in **Figure 8** represents a suitable configuration for ground resistive sensors placed in the lower positions of the left branch of the bridge. The output of the differential bridge is connected to an OpAmp-based differential amplifier with a voltage gain of A. Consequently, the single-ended output is sent to a voltage-inverting integrator whose goal is to create a stable negative feedback loop and provide a correct control voltage value (V_{CTRL}) for tunable resistors (R_{VCR}). If the measuring variation occurs to a specified range, the unbalanced output voltage is amplified, and the integrator produces a path that tracks the R_{VCR} elements until a new equilibrium condition is reached (i.e., automatic range).

Several studies have been conducted by researchers to improve the measurement accuracy of GMR sensor. Recently, Li and Dixon [27] have proposed the use of the closed-loop circuit to improve the measurement accuracy of GMR sensor. By using the biasing coil and feedback

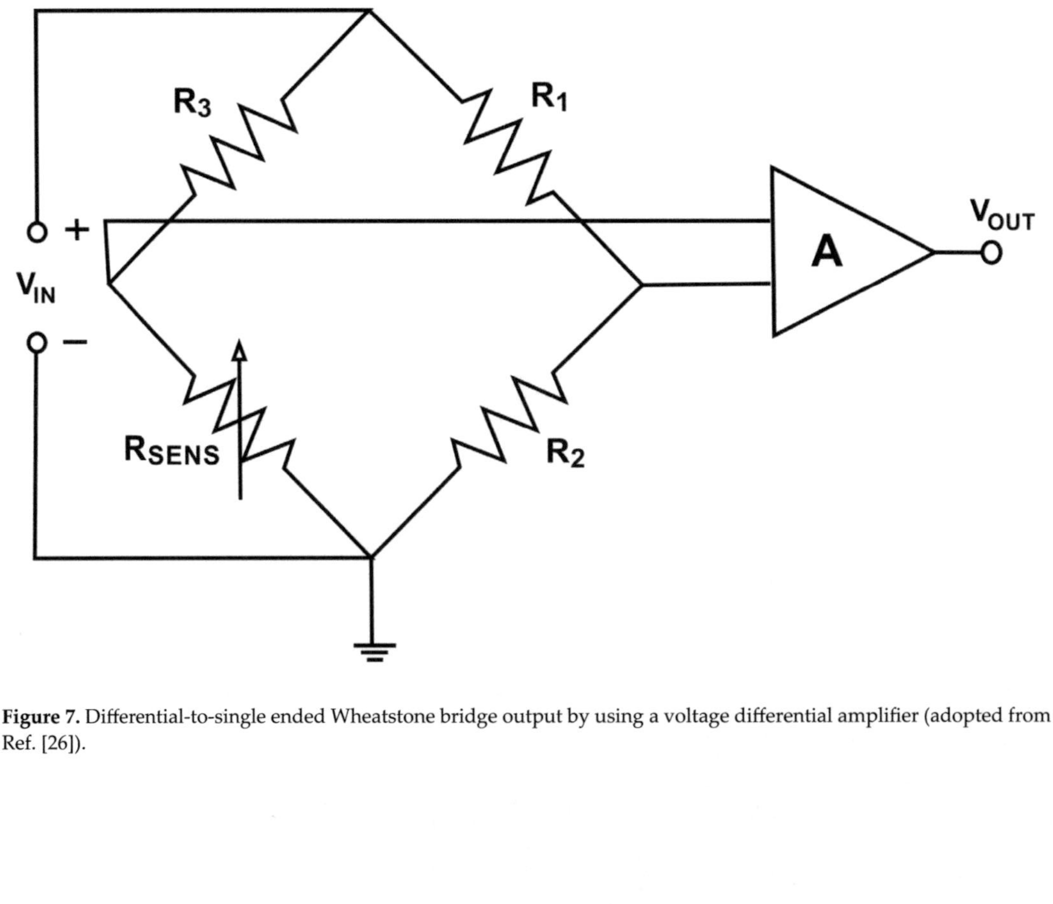

Figure 7. Differential-to-single ended Wheatstone bridge output by using a voltage differential amplifier (adopted from Ref. [26]).

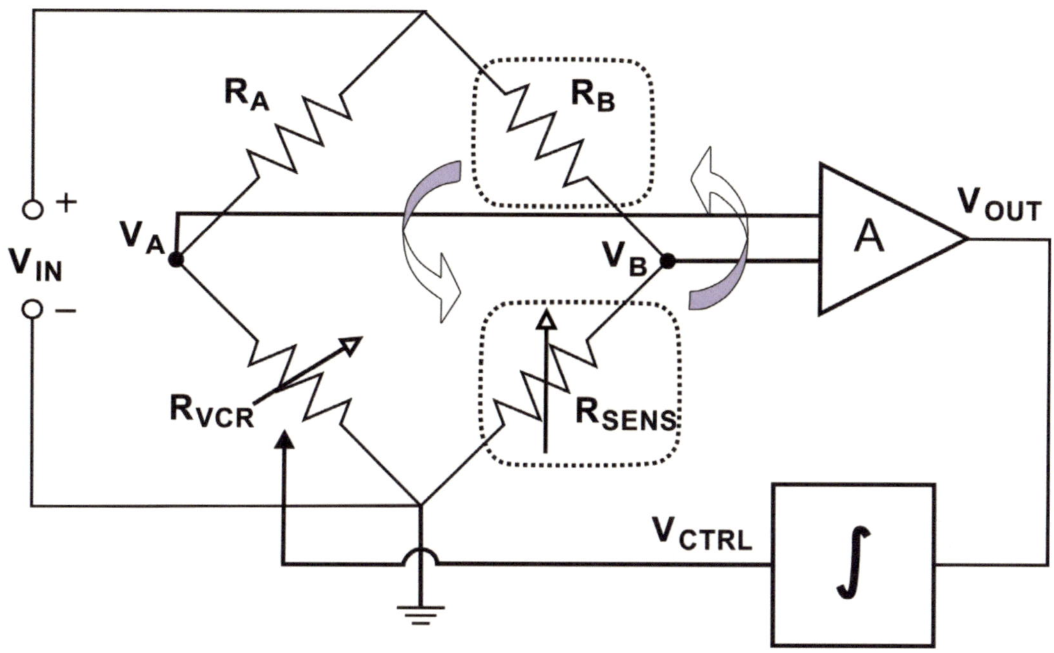

Figure 8. Block schemes of bridge-based interfaces (adopted from Ref. [26]).

circuit, the current flowing in the biasing coil is controlled by the GMR output voltage to make the GMR output constant. Hysteresis and nonlinearity of the GMR sensor have greatly minimized. Therefore, linearity and accuracy of magnetic field measurements have improved significantly.

Various efforts have been made to improve the performance of GMR sensors. An increase of 4 dB in the signal-to-noise ratio of CPP GMR sensors has been reported by Mihajlovic et al. [28]. They use Heusler alloy magnetic layers and insert an In—Zn—O electrically conductive oxide into an Ag-based metallic spacer layer.

Recently, Choi et al. [29] have shown an effective method for the enhancement of MR ratio of CPP-GMR spin-valve sensor by improving order in the B2 polycrystalline Heusler alloy films by inserting a CoFeBTa or CoBTi amorphous ferromagnetic underlayer.

5. Application of GMR sensor: recent and future trends

GMR sensors have been widely used in power systems, aerospace, modern transport systems, and the biomedical field due to the high sensitivity and wide range of magnetic field frequency response. The first application is a read head in a magnetic disk drive with the spin-valve structures. A spin-valve sensor is made of a magnetic layer exhibiting a strong coercivity (hard layer) separated by a magnetic layer with a very low coercivity (soft layer) by a thin metallic spacer.

A basic understanding of GMR device is traveling electrons from ferromagnetic layer to the other ferromagnetic layer through the conductive metal layer. If the magnetic moment in the two ferromagnetic metals is parallel then the device has a low resistance (state "0") and if the anti-parallel has high resistance (state "1"). The main attraction in the GMR technology is its ability to detect the low magnetic field. Information is stored in magnetic bits, whereas the magnetization is stored as "0" in one direction and as "1" in the other direction. This is a magnetic field detected by the GMR head. If the GMR head passes through the magnetic bits, the direction of the free layer magnetization of the head will respond with the field of each bit either spin up or down. Consequently, the magnetic moment of the free layer becomes parallel or antiparallel. This results in a change of resistance in the layers. This resistance change is detected by the GMR sensor and produces a voltage across the GMR head (while the fixed current passes through the GMR element).

The reading/writing head-integrated devices consist of a top ferromagnetic layer that is referred to as the sensing layer, and the lower ferromagnetic layer is referred to as the storage layer. The thickness of each of ferromagnetic layers is different in order to make a difference in their coercive field. The principle of reading and writing process is almost the same. However, the writing process requires a high magnetic field to rotate the moment in the storage layer.

Information is read and written by the pulse current (external magnetic field) that can detect the magnetic direction of each bit. Definition of writing is to change the magnetic moment of the storage layer (requiring high current pulses or high magnetic fields). Meanwhile, reading information on bits is changing the magnetic moment in the sensing layer (it takes a lower current pulse). The steady current passing through the GMR element is called the current sense, while the current that generates the magnetic field to rotate the magnetic moment called the word current.

The GMR sensors have been applied to measure the position of machine components as linear position detectors and transducers. A small movement of machine components (such as metal rods, gears, and other components) can produce magnetic fields. The movement along the y-axis, for example, can be determined from this magnetic field variation detected by the GMR sensor, Bx (with sensitive areas along the x-axis).

5.1. GMR sensor for detecting magnetization of liquid-contained iron

Almost all metal materials, including Fe, can be magnetized by an external magnetic field. Without an external magnetic field, the magnetic moment of the material has an irregular orientation. The magnetic moment of the material will align following the direction of the external magnetic field when the external magnetic field is applied.

The GMR sensor has been used to detect magnetic fluids containing Fe^{2+} [30] and obtain the output voltage of the GMR sensor that is proportional to the molar concentration for concentrations between 0.01 M and 0.4 M. For further experiments, it is suggested that the experiment should be conducted in a Faraday room so that the earth magnetic field cannot affect the sensor output.

More recently, GMR sensor has successfully to detect porphyrin concentration [31]. To confirm the effectiveness of this method in detecting porphyrin, we varied the flow rate and concentration of Fe^{3+}-modified porphyrin solution. **Figure 9** shows the effects of solution concentration and flow rate of the solution on the GMR output voltage. Turbidity decreases with decreasing concentration as shown in the inset image.

The result showed that the GMR sensitivity increases gradually with the increase in concentration and decrease the flow rate. Since this developed method is simple but effective for detecting porphyrin concentration, we believe that further development of this method will be a benefit for many applications, specifically relating to the medical uses.

5.2. GMR sensor for read head

Currently, we are approaching a new era of the Age of Data. Consequently, the situation changes the way we live, work, and play. That change has been initiated from the autonomous car to humanoid robots and intelligent personal assistant to the smart home devices. All of them require data storage media with very large capacity. According to a report from the International Data Corporation sponsored by Seagate, it recently states that the amount of data created worldwide will increase tenfold in 2025 [32].

In the field of digital data storage, hard disk drive (HDD) maintained its leading position among other data storage devices. In the HDD-based magnetic recording such as GMR sensors, the information is stored in the area that has been magnetized in a thin film. Transitions between similar areas are named a "bit" to be detected by a "read head" on the HDD. The number of bits per unit area is called the "areal density." Since the 1990s areal density is

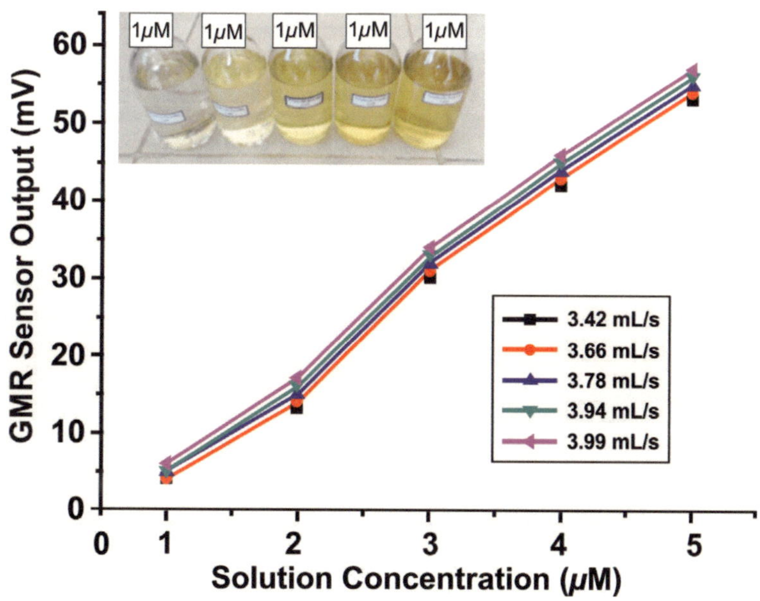

Figure 9. Effects of solution concentration and flow rate on the GMR sensor sensitivity [31].

increased dramatically, with a compound growth rate (CGR) increased to 100%, as shown in **Figure 10**. Accordingly, the need of magnetic storage media with a large capacity and small size requires a serious development of magnetic field sensors based on GMR material.

The GMR head is an analog device that detects magnetic marks with magnitude above high-resolution disks, rather than directly detecting the binary magnetization of stored bits. GMR head development trends are intended to achieve large-scale data storage densities. A simple diagram of the GMR head is shown in **Figure 11**. The sensing layer in the GMR head consists of a free and a reference layer, which is separated by a nonmagnetic layer.

Another phenomenon related to GMR is spin torque. Nowadays, a high-sensitivity spin-torque oscillator (STO) for ultrasmall field sensors has been used for storage devices. STO usually appears in one of two very different architectures: (i) nano-pillars approximately 100 nm in diameter and (ii) nano-contact, where the current enters a long magnetic structure through constriction [34].

The increase of the capacity of data storage devices raises the gap between processor speed and off-chip memory speed, resulting in increased demand for on-chip memory, more recently. The way to limit power consumption and to save memory gaps is by modifying the memory hierarchy by integrating instability at different levels, which will cause static power and also pave the way to normal-off/instant-on computation [34].

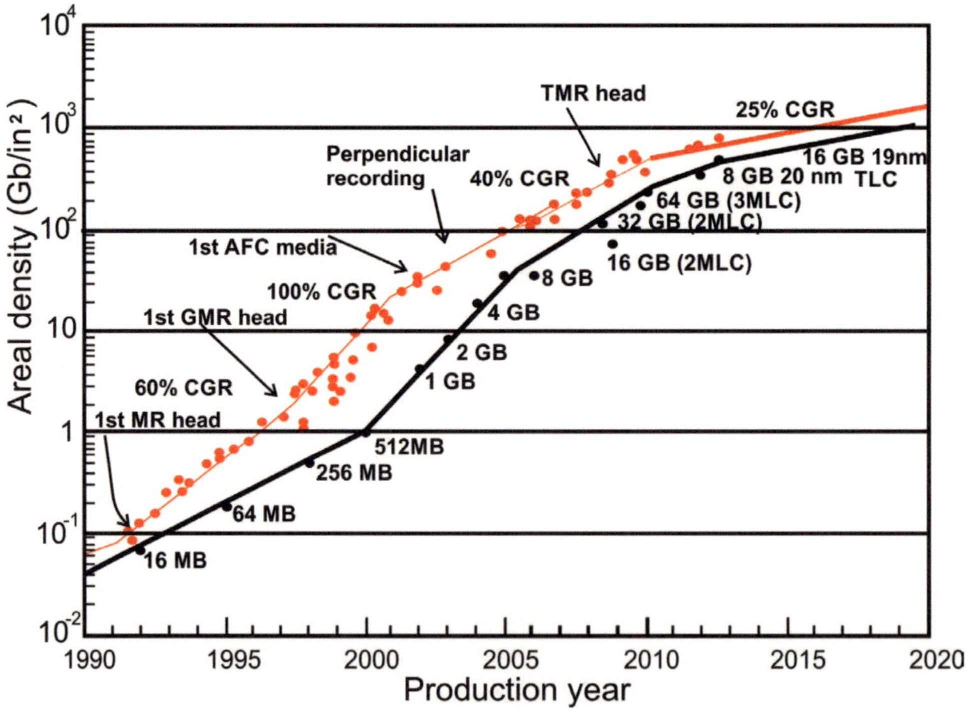

Figure 10. Trend of the increase of areal density of HDD and flash disk to the year of production [33].

5.3. GMR biosensor and biomedical application

Since the late 1990s, magnetoelectronics has emerged as one of several new platform technologies for biosensor and biochip development [36]. This technology is based on the detection of biologically functionalized micrometer- or nanometer-sized magnetic labels, using high-sensitivity microfabricated magnetic field sensors.

The development of a biosensing platform that is powerful, flexible, and high throughput is expected to have broad implications in medicine, nursing clinical diagnostics, pharmaceutical drug development, genomics, and proteomics research. It is enabled by nanotechnologies, which is emerging rapidly (i.e nanoparticles, nanotubes, and nanowires) and micro manufacturing technology (i.e MEMS- micro electro mechanical systems, microfluidics, and CMOS- complementary metal-oxide-semiconductor). Some platforms new sensing has been proposed and tested for biomedical applications, one of which is a GMR biosensor [37].

Biomolecular detection using GMR sensors is based on the principle that the resistance of the GMR sensor changed when an external magnetic field is applied. When a magnetically labeled biomolecule is brought close to the sensor, a signal will be transmitted and read by the GMR sensor. Today, GMR-based biosensors are very sensitive to detect magnetic information so that it has become a dominant player in the field of biosensors.

The use of GMR sensors for magnetic marker detection was first developed by Tondra et al. [38]. They concluded that all sizes of a single magnetic marker can be detected by the GMR sensor provided the sensor size is almost identical to the size of the marker as well as a thin isolated protective layer. Millen et al. [39] has proposed the incorporation of GMR structures on bacterial sensing as shown in **Figure 12**. Surface sensing regions of GMR need to be modified to allow for binding of antibody capture. If a sample solution containing a target antigen concerns the GMR sensor, then a complex bond exists between the target antigen and the antibody. Furthermore, it is followed by the addition of magnetic-coated antibody particles and labeled the antigen target to form a sandwich-like structure.

Currently, one of the topics of biomedical research interest is the detection of biological species. Basic research in this field will result in many immediate applications, such as food-borne

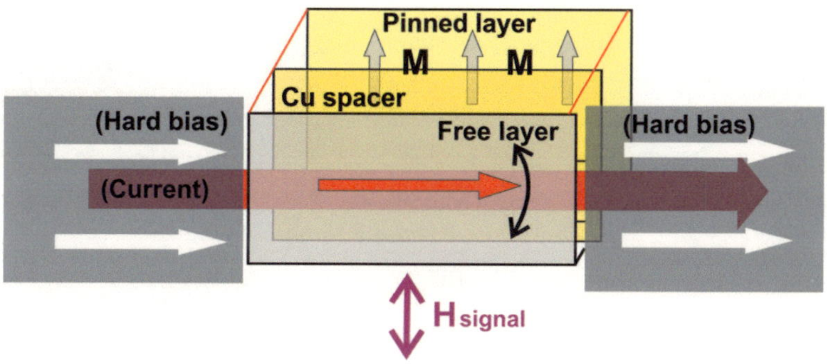

Figure 11. A simple diagram of the GMR head (adopted from Ref. [35]).

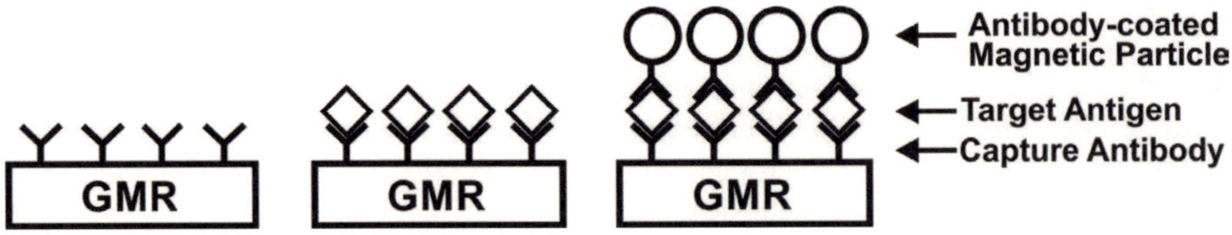

Figure 12. Bacterial detection scheme with GMR biosensor (adapted from Ref. [39]).

pathogen detection, biological war defense, or bio-diagnosis. In this scenario, it is necessary to develop an easy, inexpensive, and quick method to detect this agent. Manteca et al. [40] combined superparamagnetic particles with GMR sensors to detect target species. Recently, Elaine et al. [41] have shown examples of sensitive and specific multiplexed detection of major peanut allergens and wheat allergens gliadin using an array of GMR sensors. They found that the multiplexing capability of GMR sensor arrays provides higher levels of information that is unavailable with current commercialized enzyme-linked immunosorbent assay (ELISA) detection kits.

Krishna et al. [42] have developed a simple and sensitive method for detecting influenza A viruses using GMR biosensors. This test uses monoclonal antibodies against viral nucleoproteins (NP) in combination with magnetic nanoparticles (MNPs). The presence of influenza viruses allows the binding of MNPs to GMR sensors; consequently, their binding is proportional to viral concentrations. The binding of MNP to the GMR sensor causes a change in the sensor resistance, as measured in real-time electrical readings. Illustration of GMR biosensor for detecting influenza A virus is shown in **Figure 13**.

The GMR sensor has been used in heart rate monitoring. Kalyan et al. [43] have developed a simple cardiac rate monitoring system using GMR sensor. The GMR sensor is placed on the human wrist and provides a magneto-plethysmographic signal. This signal is processed by a simple analog and digital instrumentation stage to give an indication of heart rate. The prototype of this system is shown in **Figure 14**.

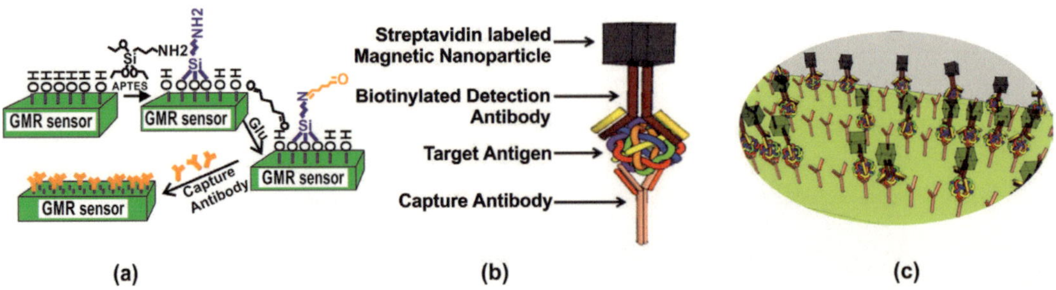

Figure 13. Schematic representation of GMR biosensor. (a) Schematic diagram of GMR biosensor surface functionalization. (b) Schematic drawing of a typical sandwich structure (biotinylated detection antibody/target antigen/capture antibody). (c) Schematic illustration of influenza A virus detection (adopted from Ref. [42]).

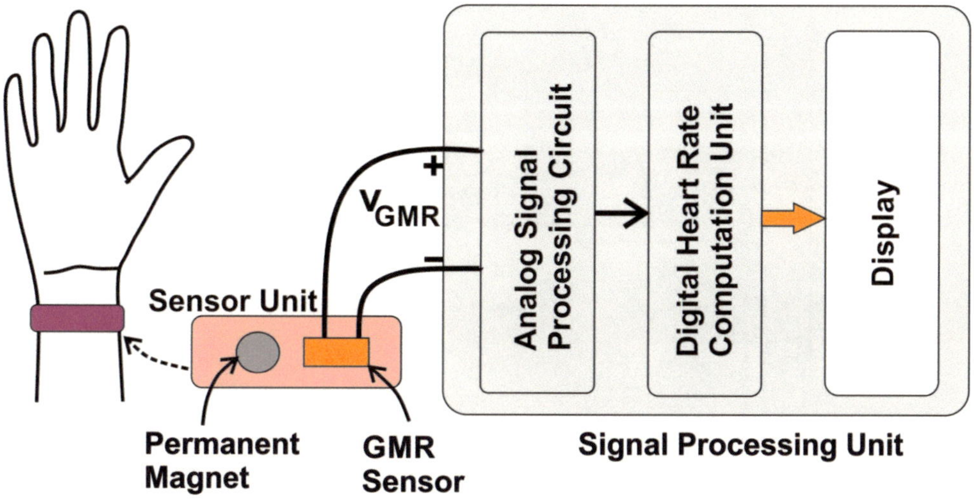

Figure 14. Simplified diagram of the proposed GMR heart rate monitor. It uses a GMR sensor and magnet in the sensor unit to get the plethysmograph (v_{GMR}) and heart rate (adopted from Ref. [43]).

5.4. GMR sensor as magnetometer

The GMR sensors have advantages of low power consumption, low cost, high detection capability, linearity, and three-dimensional (3D) measurement capabilities, thus matching the need for a magnetometer. Luong et al. [44] have proposed a three-dimensional GMR sensor design with a single bridge and one flux guide for the three-dimensional magnetometer. This design helps to reduce the sensor size, power consumption, and fabrication cost.

Recently, Xiao [45] has reported the use of GMR sensors for steering wheel angle sensors with high accuracy, wide measurement range, and simple structure. It used two GMR chips to detect magnetic fields. The GMR chip measures the rotation angle of the multipole magnetic ring with 120 couples of a magnetic pole, whereas each magnetic pole of which outputs an angle signal of 0–360.

In nondestructive test (NDT), GMR sensor as a magnetometer has implemented to detect the material defect profile. The sensing axis of the GMR sensor is set perpendicular to the direction of the excitation magnetic field; consequently, the information collected primarily reflects changes in the eddy current caused by the defect. Moreover, application of the GMR sensor as a signal receiver has increased the sensitivity of this technique in the detection of small defects [46]. The implementation of the GMR sensor on NDT eddy current techniques is still new, and many issues related to hybrid systems (coils-GMR) will continue to be explored by researchers.

Recently, Gao et al. [47] have designed a 4-GMR probe with a rectangular coil as the excitation coil and an eddy current detection system to detect weld defect (**Figure 15**). The experiments conducted by Gao et al. [47] showed that through the method of the proposed detection system, the recognition rate was 92% for flawless welding and 90% for welding with defects, with an overall recognition rate of 90.9%. This shows that the method can detect weld defects effectively.

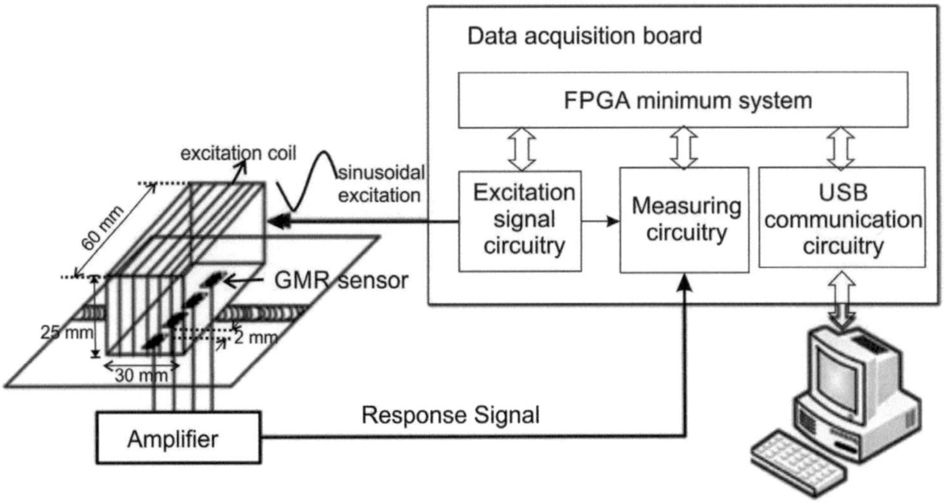

Figure 15. Schematic graph of eddy current testing system and weld inspection based on GMR sensor (adopted from Ref. [47]).

6. Conclusion

The use of GMR material for sensor applications is increasing rapidly, as a result of its small size, high signal level, high sensitivity, high-temperature stability, low power consumption, low cost, and compatibility with CMOS electronics. Some sensor applications have been developed using GMR material, e.g., magnetic sensors, read head sensors, biosensors, and many others. In the future, the development of GMR-based sensors will increase continuously, especially for health application, data storage, and daily needs.

Acknowledgements

This work was partially supported by Research Incentive on National Innovation System (SINas), Ministry of Research, Technology, and Higher Education of the Republic of Indonesia, Research Code: RD-2015-0352.

Author details

Mitra Djamal[1,2*] and Ramli Ramli[3]

*Address all correspondence to: mitra.djamal@yahoo.co.id

1 Department of Physics, Institut Teknologi Bandung, Bandung, Indonesia

2 Department of Physics, Institut Teknologi Sumatera, Lampung Selatan, Indonesia

3 Universitas Negeri Padang, Padang, Indonesia

References

[1] IC Insight. Optoelectronics, Sensors/Actuators, and Discretes Move toward Normal Growth Rates [Internet]. 2017. Available from: www.icinsights.com [Accessed: April 20, 2017]

[2] Traenkler H-R, Kanoun O. Smart system and device: Innovative key modules for engineering application. In: Derbel F. editor. Smart Systems and Devices; March 27-30; Hammamet. Tunisia: SOGIC; 2001. pp. 3-12 .

[3] Djamal M, Ramli. Development of sensors based on giant magnetoresistance material. Procedia Engineering. 2012;**32**:60-68. DOI: 10.1016/j.proeng.2012.01.1237

[4] Djamal M, Ramli, Khairurrijal, Haryanto F. Development of giant magnetoresistance material based on cobalt ferrite. Acta Physica Polonica A. 2015;**128**(2B):19-22. DOI: 10.12693/APhysPolA.128.B-19

[5] Lenz J, Edelstein AS. Magnetic sensors and their applications. IEEE Sensors Journal. 2006;**6**(3):631-649. DOI: 10.1109/JSEN.2006.874493

[6] Herrera-May AL, Aguilera-Cortés LA, García-Ramírez PJ, Manjarrez E. Resonant magnetic field sensors based on MEMS technology. Sensors. 2009;**9**:7785-7813. DOI: 10.3390/s91007785

[7] Fert A. Nobel lecture: Origin, development, and future of spintronic. Reviews of Modern Physics. 2008;**80**(4):1517-1530. DOI: 10.1103/RevModPhys.80.1517

[8] Grünberg PA, Takanashi K. Spintronics: Towards devices with lower energy consumption. In: 10th IEEE International Conference on Nanotechnology; August 17-20; Seoul. South Korea: IEEE; 2010. p. 3. DOI: 10.1109/NANO.2010.5698059

[9] Sun D, Ehrenfreund E, Vardeny ZV. The first decade of organic Spintronic research. Chemical Communications. 2014;**50**:1781-1793. DOI: 10.1039/C3CC47126H

[10] Djamal M, Ramli. Thin film of Giant Magnetoresistance (GMR) material prepared by sputtering method. Advanced Materials Research. 2013;**770**:1-9. DOI: 10.4028/www.scientific.net/AMR.770.1

[11] Tezuka N. New materials research for high spin polarized current. Journal of Magnetism and Magnetic Materials. 2012;**324**:3588-3592. DOI: 10.1016/j.jmmm.2012.02.097

[12] Moussy JP. From epitaxial growth of ferrite thin films to spin-polarized tunnelling. Journal of Physics D: Applied Physics. 2013;**46**(14):143001. DOI: 10.1088/0022-3727/46/14/143001

[13] Goto K, Kanki T, Kawai T, Tanaka H. Giant magnetoresistance observed in $(Fe,Mn)_3O_4$ artificial Nanoconstrained structures at room temperature. Nano Letters. 2010;**10**(8):2772-2776. DOI: 10.1021/nl100542a

[14] Culity BD, Graham CD. Introduction to Magnetic Materials. 2nd ed. New York: John Wiley & Sons, Inc; 2008. p. 568

[15] Reddy KM, Padture NP, Punnoose A, Hanna C. Magnetoresistance characteristics in individual Fe_3O_4 single crystal nanowire. Journal of Applied Physics. 2015;**117**:17E115. DOI: 10.1063/1.4914535

[16] Dho J, Kim B-G, Ki S. Distinctive uniaxial magnetic anisotropy and positive magnetoresistance in (110)-oriented Fe_3O_4 films. Journal of Applied Physics. 2015;**117**:163904. DOI: 10.1063/1.4918915

[17] Liu E, et al. Inverse magnetoresistance in single layer Fe_3O_4 film. In: 2015 IEEE International Magnetics Conference (INTERMAG); 11-15 May; Beijing. China: IEEE; 2015. p. 1. DOI: 10.1109/INTMAG.2015.7157433

[18] Huang ZC, Yue JJ, Wang J, Zhai Y, Xu YB, Wang BP. Oscillatory tunneling Magnetoresistance in Fe_3O_4/n-GaAs/Fe_3O_4 junction. IEEE Transactions on Magnetics. 2015;**51**(11):1300604. DOI: 10.1109/TMAG.2015.2435038

[19] Mitra A, Barick B, Mohapatra J, Sharma H, Meena SS, Aslam M. Large tunneling magnetoresistance in octahedral Fe_3O_4 nanoparticles. AIP Advances. 2016;**6**:055007. DOI: 10.1063/1.4948798

[20] Ramli et al. Novel ternary $CoFe_2O_4$/CuO/$CoFe_2O_4$ as a giant magnetoresistance sensor. Journal of Mathematical and Fundamental Sciences. 2016;**48**(3):230-240. DOI: 10.5614/j.math.fund.sci.2016.48.3.4

[21] Tumanski S. Thin Film Magnetoresistive Sensors. Florida: CRC Press; 2001 576 p

[22] NVE Corporation. GMR Sensors Catalog [Internet]. Available from: http://www.nve.com/sensorcatalog.php [Accessed: April 10, 2017]

[23] Reig C, Beltran MDC, Munoz DR. Magnetic field sensors based on giant magnetoresistance (GMR) technology: Applications in electrical current sensing. Sensors. 2009;**9**:7919-7942. DOI: 10.3390/s91007919

[24] Nor AFM, Hill EW, Parker MR. Geometry effects on low frequency noise in giant magnetoresistance sensors. IEEE Transaction on Magnetic. 1998;**34**(4):1327-1329. DOI: 10.1109/20.706537

[25] Nordling J, Millen RL, Bullen HA, Porter MD, Tondra M, Granger A. Giant magnetoresistance sensors. 1. Internally calibrated readout of scanned magnetic arrays. Analytical Chemistry. 2008;**80**(21):7930-7939. DOI: 10.1021/ac8009577

[26] Reig C, de Freitas SC, Mukhopadhyay SC. Giant Magnetoresistance (GMR) Sensors. 1st ed. Berlin: Springer-Verlag; 2013. p. 301. DOI: 10.1007/978-3-642-37172-1

[27] Li Z, Dixon S. A closed-loop operation to improve GMR sensor accuracy. IEEE Sensors Journal. 2016;**16**(15):6003-6007. DOI: 10.1109/JSEN.2016.2580742

[28] Mihajlovic G, Nakatani TM, Smith N, Read JC, Choi Y-S, Tseng H-W, Childress JR. Enhanced signal-to-noise ratio in current-perpendicular-to-plane giant-magnetoresistance sensors by suppression of spin-torque effects. IEEE MAGNETICS LETTERS. 2015;**6**:3001104. DOI: 10.1109/LMAG.2015.2496552

[29] Choi Y-S, Nakatani T, Read JC, Carey MJ, Stewart DA, Childress JR. Enhancement of current-perpendicular-to-plane giant magnetoresistance by insertion of amorphous ferromagnetic underlayer in Heusler alloy-based spin-valve structures. Applied Physics Express. 2017;**10**(1):013006. DOI: 10.7567/APEX.10.013006

[30] Ramli, Muhtadi AH, Sahdan MF, Haryanto F, Khairurrijal, Djamal M. The preliminary study of giant magnetoresistance sensor for detection of oxygen in Human's blood. AIP Conference Proceedings. 2009;**1325**:309-312. DOI: 10.1063/1.3537937

[31] Aminudin A, Tjahyono DH, Suprijadi, Djamal M, Zaen R, Nandiyanto ABD. Solution concentration and flow rate of Fe^{3+}-modified porphyrin (red blood model) on giant magnetoresistance (GMR) sensor efficiency. IOP Conference Series: Materials Science and Engineering. 2017;**180**(1):012137. DOI: 10.1088/1757-899X/180/1/012137

[32] Reinsel D, Gantz J, Rydning J. Data Age 2025: The Evolution of Data to Life-Critical [Internet]. 2017. Available from: www.idc.com [Accessed: April 12, 2017]

[33] Marchon B, Pitchford T, Hsia Y-T, Gangopadhyay S. The head-disk interface roadmap to an areal density of 4 Tbit/in2. Advances in Tribology. 2013;**2013**:521086. DOI: 10.1155/2013/521086

[34] Stamps RL et al. The 2014 magnetism roadmap. Journal of Physics D: Applied Physics. 2014;**47**(33):333001. DOI: 10.1088/0022-3727/47/33/333001

[35] Fullerton EE, Childress JR. Spintronics, magnetoresistive heads, and the emergence of the digital world. Proceedings of the IEEE. 2016;**104**(10):1-9

[36] Xu L, Yu H, Michael SA, Han SJ, Osterfeld S, White RL, Pourmand N, Wang SX. Giant magnetoresistive biochip for DNA detection and HPV genotyping. Biosensors and Bioelectronics. 2008;**24**(1):99-103. DOI: 10.1016/j.bios.2008.03.030

[37] Djamal M, Ramli, Haryanto F, Khairurrijal. GMR biosensor for clinical diagnostic. In: Serra PA, editor. Biosensors for Health, Environment and Biosecurity. 1st ed. Rijeka: IntechOpen; 2011. p. 149-164. DOI: 10.5772/16365

[38] Tondra M, Porter M, Lipert RJ. Model for detection of immobilized superparamagnetic nanosphere assays labels using giant magnetoresistive sensors. Journal of Vacuum Science & Technology A: Vacuum, Surfaces, and Films. 2000;**18**:1125. DOI: 10.1116/1.582476

[39] Millen RL, Kawaguchi T, Granger MC, Porter MD. Giant magnetoresistive sensors and Superparamagnetic nanoparticles: A chip-scale detection strategy for immunosorbent assays. Analitical Chemistry. 2005;**77**(20):6581-6587. DOI: 10.1021/ac0509049

[40] Mantecaa A, Mujikab M, Arana S. GMR sensors: Magnetoresistive behaviour optimization for biological detection by means of superparamagnetic nanoparticles. Biosensors and Bioelectronics. 2011;**26**(8):3705-3709. DOI: 10.1016/j.bios.2011.02.013

[41] Ng E, Nadeau KC, Wang SX. Giant magnetoresistive sensor array for sensitive and specific multiplexed food allergen detection. Biosensors and Bioelectronics. 2016;**80**:359-365. DOI: 10.1016/j.bios.2016.02.002

[42] Krishna VD, Wu K, Perez AM, Wang J-P. Giant magnetoresistance-based biosensor for detection of influenza A virus. Frontiers in Microbiology. 2016;**7**:PMC4809872. DOI: 10.3389/fmicb.2016.00400

[43] Kalyan K, Chugh VK, Anoop CS. Non-invasive heart rate monitoring system using giant magneto resistance sensor. In: 2016 IEEE 38th Annual International Conference of the Engineering in Medicine and Biology Society (EMBC), 16-20 Aug. 2016, Orlando. USA: IEEE; 2016. p. 16395106 . DOI: 10.1109/EMBC.2016.7591819

[44] Luong V, Jeng J, Lai B, Hsu J, Chang C, Lu C. Design of three-dimensional magnetic field sensor with single bridge of spin-valve giant magnetoresistance films. In: 2015 IEEE International Magnetics Conference (INTERMAG), 11-15 May, Beijing. China: IEEE; 2015. p. 15303793. DOI: 10.1109/INTMAG.2015.7156762

[45] Xiao M. A high-accuracy steering wheel angle sensor based on GMR. In: 2016 Sixth International Conference on Instrumentation & Measurement, Computer, Communication and Control (IMCCC), 21-23 July 2016, Harbin. China: IEEE; 2016. pp. 1-4. DOI: 10.1109/IMCCC.2016.115

[46] Poon TY, Tse NCF, Lau RWH. Extending the GMR current measurement range with a counteracting magnetic field. Sensors. 2013;**13**:8042-8059. DOI: 10.3390/s130608042

[47] Gao P, Wang C, Li Y, Wang L, Cong Z, Zhi Y. GMR-based eddy current probe for weld seam inspection and its non-scanning detection study. Nondestructive Testing and Evaluation. 2017;**32**(2):133-151. DOI: 10.1080/10589759.2016.1149583

Chapter 7

AMR Sensor and its Application on Nondestructive Evaluation

Dongfeng He

Additional information is available at the end of the chapter

http://dx.doi.org/10.5772/intechopen.70334

Abstract

To improve the performance of anisotropic magnetoresistance (AMR) sensor, a low-noise driving circuit for the AMR sensor was developed and the magnetic field noise spectral densities of 12 pT/root(Hz) at 1 kHz and 20 pT/root(Hz) at 100 Hz were achieved. The driving circuit could operate in amplifier mode or feedback mode. For the driving circuit with feedback, the distortion of the system was reduced and the AMR sensor was suitable for the applications in the environment without shielding. The Set/Reset method was used to reduce the low frequency noise of the AMR sensor. Due to the low noise of AMR sensor, the eddy current testing (ECT) system with the AMR sensor had the advantage of detecting deep and small defects in metal structures. The dual frequency ECT system was developed to reduce the influence of lift-off variance. Using the ECT system with the AMR sensor, we successfully detected the small defects in the combustion chamber of liquid rocket.

Keywords: AMR (anisotropic magneto-resistance) sensor, NDE (nondestructive evaluation), ECT (eddy current testing), driving circuit, defect, combustion chamber

1. Introduction

Eddy current testing (ECT) is an effective nondestructive evaluation (NDE) method to detect defects in metal (or conductive) structures. For the ECT NDE method, alternating magnetic field is produced when alternating current (AC) current flows in the excitation coil; so eddy current is induced in the conductive sample and the defect in the sample changes the amplitude and the distribution of the eddy current. By measuring the change of the eddy current, the defect can be detected.

Many magnetic sensors have been used to construct ECT NDE systems, such as inductive coil [1], hall sensor [2], giant magnetoresistive (GMR) sensor [3, 4], anisotropic magnetoresistance

(AMR) sensor [5, 6], flux gate [7, 8] and superconducting quantum interference device (SQUID) [9–11]. Compared with other sensors, SQUID has the lowest noise, but it needs liquid nitrogen or liquid helium for the cooling, which is not convenient for some industrial applications. Stutzke et al. [12] measured the noise spectral densities in the frequency range from 0.1 Hz to 10 kHz on a variety of commercially available magnetic sensors of GMR sensors, AMR sensors and tunnel magnetoresistance (TMR) sensors. The results show that AMR sensor had the lowest magnetic field noise spectrum density with the order of 100 pT/root(Hz) in the frequency range from 1 Hz to 1 kHz. And we also achieved much better performances by optimizing the driving circuit of the AMR sensor [13], so we chose AMR sensor to construct our ECT NDE systems.

In this chapter, we will summarize our research on AMR sensor and its application on ECT NDE. Firstly, we will give an introduction of AMR sensor; secondly, describe the optimization of the AMR sensor: lowering the noise, feedback operation and reducing the temperature effect using Set/Reset method; then, the application of AMR sensor to NDE for the defect detection of the combustion chamber of liquid rocket; finally, the conclusion.

The AMR effect was first discovered in 1857 by Thomson [14]. Different from the ordinary magneto-resistive effect in metals, the AMR effect appeared in ferromagnetic materials [15], which was explained as the spin-orbit interaction. In this effect, the resistivity depended on the orientation of magnetization with respect to the direction of the electric current. The change of the longitudinal resistivity of the material magnetized parallel to the current direction was larger than that of the transverse resistivity of the material magnetized perpendicular to the current direction [15]. In the absence of the external magnetic field, the direction of the magnetization was along the anisotropy axis (the so-called easy axis). To change the direction of the magnetization, the applied magnetic field should be perpendicular to the anisotropy axis, so the direction perpendicular to the anisotropy axis was called the sensitive axis. The thin film of the NiFe alloy (permalloy) was most frequently used for preparing AMR sensors. **Figure 1** shows the typical magnetoresistive coefficient ΔR/R for a thin film permalloy magnetoresistor.

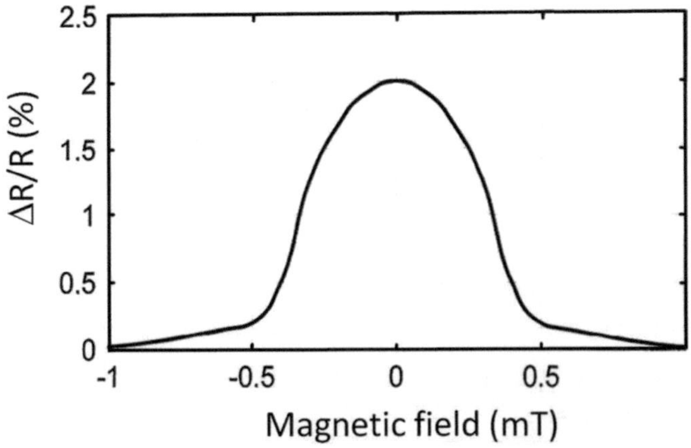

Figure 1. Typical magnetoresistive coefficient of an AMR thin film permalloy magnetoresistor.

To obtain the linear characteristics of an AMR sensor, the current path should be 45° inclined to the anisotropy axis [16], as shown in **Figure 2**.

The resistive Wheatstone bridge was often used to construct the AMR sensor, such as the HMC1001 of Honeywell [17]. For the AMR sensor with Wheatstone bridge, it only required a supply voltage to bias the sensor. **Figure 3(a)** shows the Wheatstone bridge element of an AMR sensor. The sensitive directions of magnetoresistor 1 and 3 were opposite to the sensitive directions of magnetoresistor 2 and 4. For some commercial AMR sensors, Offset straps and Set/Reset straps were often integrated with the AMR sensor chips. **Figure 3(b)** shows it.

The Offset strap was a spiral of metallization that coupled to the sensor element's sensitive axis. Using the Offset strap, the unwanted signal could be compensated, such as the offset voltage of the bridge and the DC magnetic field. Another advantage of the Offset strap was that it could be used to construct the feedback circuit of the AMR sensor and the distortion of output signal could be reduced.

When AMR sensor was exposed in a strong magnetic field (>4–20 Gauss), which could demagnetize the AMR sensor [18], the AMR effect was reduced and the sensitivity of the AMR sensor could be significantly reduced. To recover the sensitivity of the AMR sensor, the Set/Reset

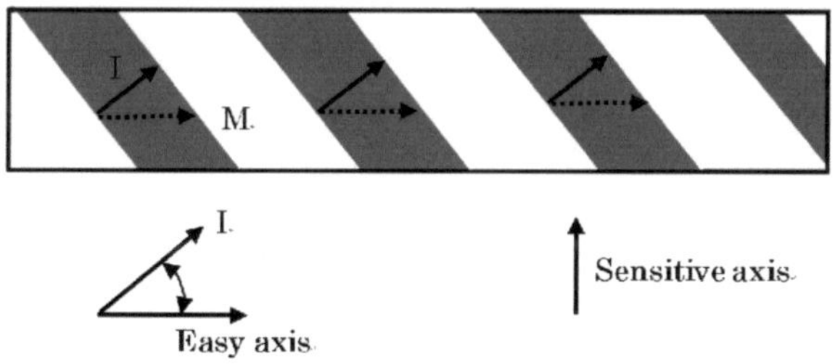

Figure 2. AMR sensor with the inclination of the current path by the angel 45° to the easy axis.

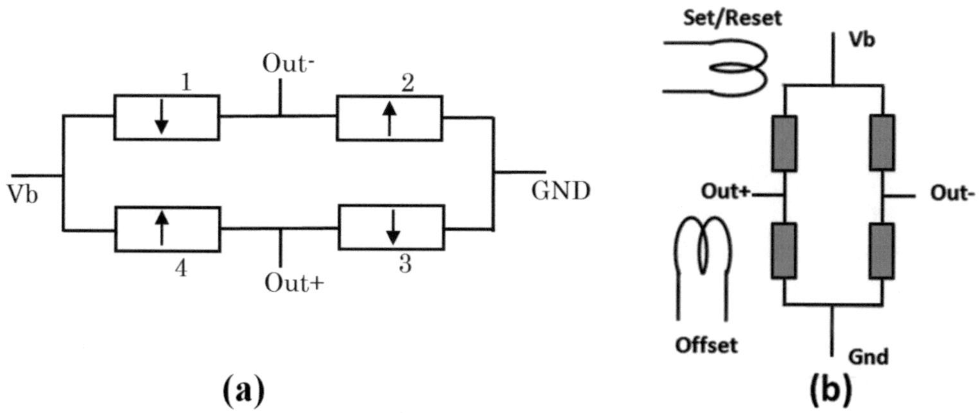

Figure 3. (a) Magnetoresistive Wheatstone bridge elements of AMR sensor. (b) The construction of AMR sensor with Offset strap and Set/Reset strap.

strap was used, which was another spiral of metallization that coupled to the sensor element's easy axis (perpendicular to the sensitive axis on the sensor die). When a high current pulse was applied to the Set/Reset strap, the magnetization of the AMR sensor was recovered and the AMR sensor was Set/Reset to high sensitivity. The sensitive direction was reversed when the reset current direction was opposite. Using the Set/Reset strap, we developed a reversal bias method to reduce the influence of temperature drift to the AMR sensor, which will be described in Section 2.

Table 1 shows the typical performance data of the HMC1001 AMR sensor of Honeywell. To understand the properties of the sensor, we give some explanations about the specifications of the AMR sensor. The sensitivity (S) of a magnetic sensor means the output voltage of the sensor when one Gauss magnetic field is applied. The typical sensitivity of the AMR sensor HMC1001 is 3.2 mV/V/Gauss. For the AMR sensor, the sensitivity has relation with the bias voltage. If the bias voltage is 5 V, the sensitivity is 3.2 × 5 = 16 mV/Gauss; if the bias voltage is 10 V, the sensitivity is 3.2 × 10 = 32 mV/Gauss.

For an AMR sensor with driving circuit, the signal is amplified (amplifier mode) or changed (feedback mode). We use field/voltage transfer coefficient (ΔB/ΔV) to express it, which means the corresponding magnetic field for 1 V output signal. For a sensor works in amplifier mode, if the total gain of the amplifier is G, we have ΔB/ΔV = G/S.

Using spectrum analyzer, we can measure the output voltage noise spectral density V_n of the sensor. Then we can get the magnetic field noise spectral density $B_n = V_n*(\Delta B/\Delta V)$.

Bridge supply voltage (V)	5
Resistance (Ω)	600–1200
Sensitivity (mV/V/Gauss)	3.2
Field range (Gauss)	−2~2
Operating temperature (°C)	−55 to 150
Linearity error at ±1 Gauss (%FS)	0.1
Linearity error at ±2 Gauss (%FS)	1
Equivalent input voltage noise density at 1 Hz (nv/√Hz)	29
Bridge offset (mv)	−60 to 30
Magnetic field noise by 10 Hz bandwidth (μGauss)	27*
Set/Reset Current (A)	2~5
Offset temperature drift without Set/Reset (%/°C)	±0.03
Offset temperature drift with Set/Reset (%/°C)	±0.001
Offset constant (mA/Gauss)	51
Bandwidth (MHz)	5

*27 is the root mean square (RMS) value.

Table 1. The specifications of the AMR sensor HMC1001 [17].

V_n has relation to the gain of the driving circuit, but B_n has no relation to the gain of the driving circuit. B_n is more commonly used to express the characteristics of a magnetic sensor. If the signal bandwidth is Δf, then the magnetic resolution is $B_n*(\Delta f)^{1/2}$. If Δf is 1 Hz, the magnetic field resolution is equal to B_n, so sometimes, we also call B_n the magnetic field resolution.

Due to its small size, low cost and high sensitivity, anisotropic magnetoresistance (AMR) sensor has been used for many applications, such as navigation, nondestructive evaluation (NDE) and vehicle detection. To improve the performance of the AMR sensor, we optimized the driving circuit of the AMR sensor. Lower noise was obtained. The distortion was reduced using feedback method, and the influence of temperature fluctuation was reduced using Set/Reset method.

2. Optimization of the AMR sensor: low noise, feedback and Set/Reset

We developed a low noise driving circuit for the AMR sensor. **Figure 4** shows the block diagram of the circuit. The driving circuit could operate in amplifier mode or feedback mode [13]. We used a voltage source to bias the AMR sensor. The output of the Wheatstone bridge of the AMR sensor was sent to a differential preamplifier made by low noise operational amplifier of LT1028. The total gain of the preamplifier and the amplifier was 500. The switch SW1 was used to change the operation mode. If SW1 was turned to OFF, it was in amplifier mode, and the output of the amplifier V_{amp} measured the magnetic field. If SW1 was turned to ON, the circuit was in feedback mode. For the feedback mode, an integrator and a feedback resistance R_f were used. Through the feedback resistance R_f, the feedback current was sent to the Offset strap of HMC1001, the signal of V_{out} measured the magnetic field.

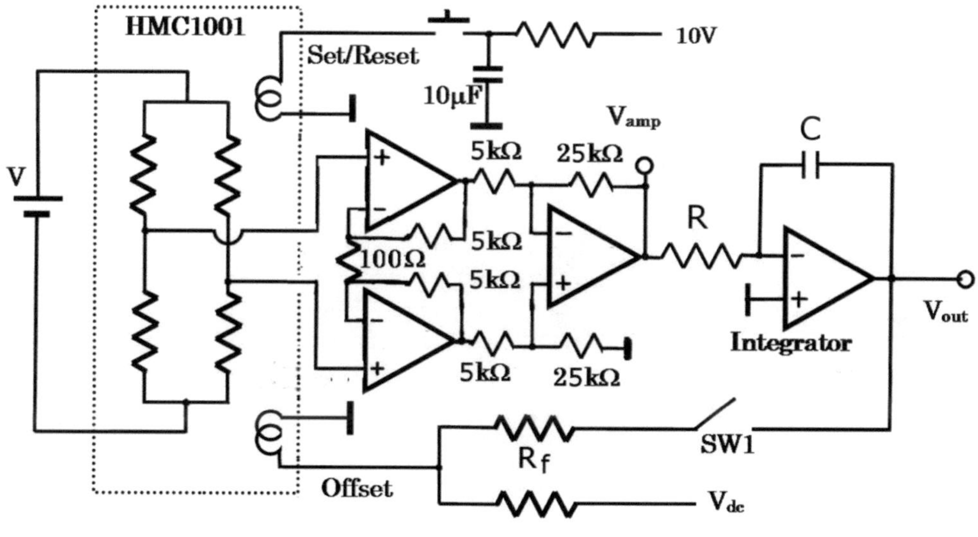

Figure 4. Block diagram of the driving circuit for AMR sensor.

The response bandwidth of the driving circuit was determined by the feedback resistance R_f, the resistance R and the capacitor C of the integrator. When R_f of 500 Ω, R of 2 kΩ and C of 220 pF were used, the response bandwidth was about 300 kHz. The offset magnetic field was adjusted by the DC voltage of V_{dc}. When the driving circuit of AMR sensor operated in feedback mode, the operation point was locked to a fixed point, thus, the distortion of the output signal was reduced.

To restore the high sensitivity of the AMR sensor, a simple Set/Reset method was used in our circuit. When the switch SW2 was OFF, the capacitor of 10 μF was charged to 10 V through the resistance of 10 kΩ. When switch SW2 was momentarily closed, a Set pulse with current amplitude of about 4 A was produced to restore the AMR sensor.

The sensitivity of the AMR sensor was proportional to the bias voltage. From the data sheet of HMC1001, the typical bias voltage was 5V and the maximum bias voltage was about 12 V. Our experiments proved that the bias voltage could be increased to 24 V and the AMR sensor still was not damaged. There were three contributions to the total noise of the AMR sensor: the noise produced by the driving circuit; the thermal noise of the resistance of the bridge; and the intrinsic magnetic noise of the AMR sensor itself. **Figure 5** shows the equivalent input voltage noise spectral density for different bias voltages from 4 to 24 V. The noise spectral density was measured when the driving circuit was at amplifier mode. For higher frequency (above 1 kHz), the thermal noise of the bridge had a big contribution to the total noise. The white noise increased a little bit for higher bias voltage. The reason was that the temperature

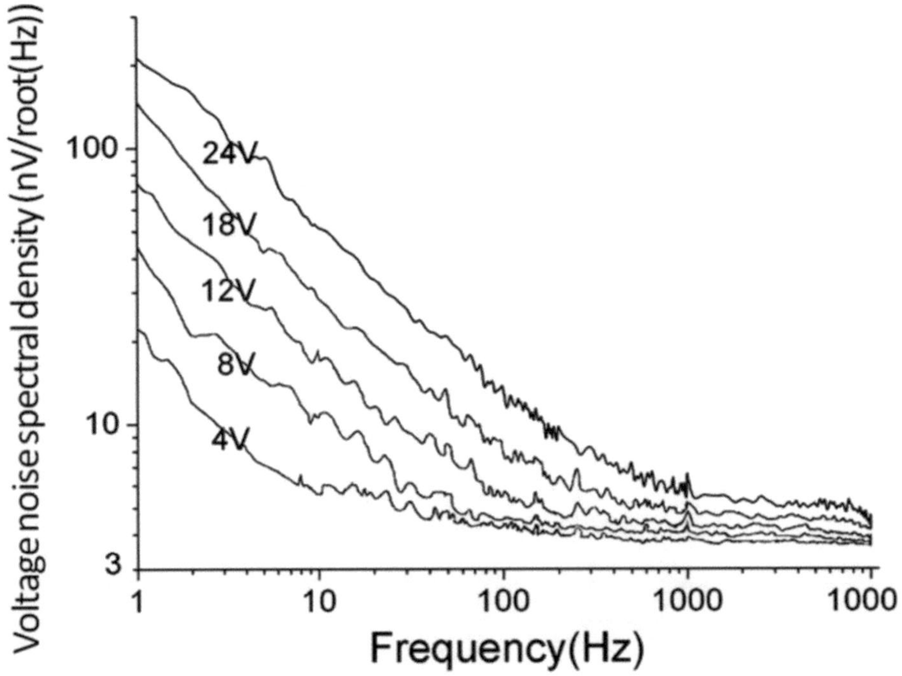

Figure 5. The equivalent input voltage noise spectral density for the bias voltage from 4 to 24 V.

of the sensor chip was increased with higher bias voltage. For lower frequency, the noise of the AMR sensor was mainly determined by the intrinsic magnetic noise of the AMR sensor, which might be caused by the random movement of the magnetic domain in the sensor film.

When the driving circuit was in feedback mode, to calibrate the AMR sensor's output, a 10 turn circular coil with a diameter of 50 cm was used to produce a magnetic field when an AC current of 230 Hz was applied to the coil. A Gauss meter and the AMR sensor were put in the center of the coil to measure the magnetic field. After calibration, the transfer coefficient of the magnetic field and the output voltage of the AMR sensor ($\Delta B/\Delta V$) was about 0.039 Gauss/V. Now, we made a simple estimation of the transfer coefficient. Since the feedback resistance was 500 Ω, the feedback current for 1 V output was 1000/500 = 2 mA/1V. Considering the offset constant (**Table 1**) was about 51 mA/Gauss, we could calculate the transfer coefficient 2/51 = 0.0392 Gauss/V, which was very close to the measured value of 0.039 Gauss/V. **Figure 6** shows the magnetic field noise spectral density of the AMR sensor for different bias voltages. The measurement was done in a magnetic shielding room. The low frequency magnetic field noise spectral densities were almost same for all bias voltages. The reason was that the intrinsic magnetic noise of the AMR sensor determined the total noise. At high frequency, the AMR sensor had lower magnetic noise spectral density when bigger bias voltage was used. When the AMR sensor had the bias voltage of 24 V, the magnetic field noise spectral densities were about 12 pT/root(Hz) at 1 kHz and 20 pT/root(Hz) at 100 Hz. These values

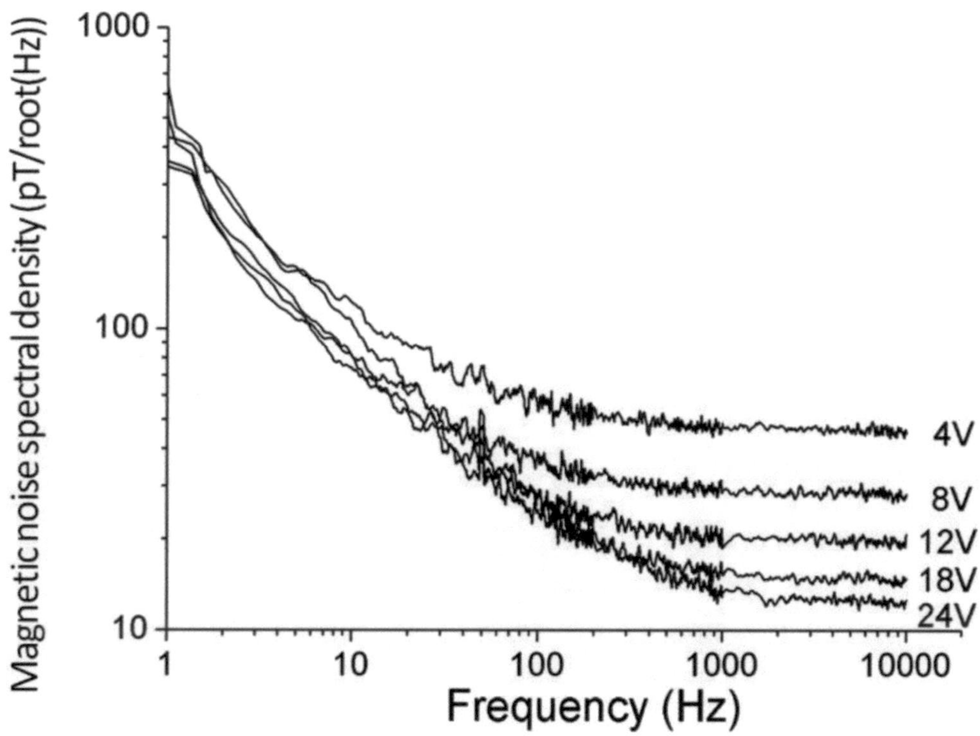

Figure 6. The magnetic field resolution of AMR sensor.

were obtained when the driving circuit was at feedback mode. We could estimate it using the sensitivity and the equivalent input noise spectral density of the AMR sensor. When the bias voltage was 24 V, the sensitivity of the AMR sensor was about 3.2 × 24 = 76.8 mV/Gauss = 0.768 nV/pT. From **Figure 5**, the equivalent input voltage noise spectral density was about 7 nV/root(Hz) at 1 kHz; then we could estimate the magnetic field noise spectrum was about 7/0.768 = 9 pT/√Hz, which was close to the measured value of 12 pT/√Hz.

When AMR sensor was biased by a voltage, the power produced by the AMR sensor made the temperature of the AMR sensor higher than environmental temperature. When a sample was close to the AMR sensor, it caused a small thermal disturbance and the AMR sensor had a signal output. We observed this phenomenon [19] and developed a method to reduce the influence of this heat transfer effect [20].

Figure 7 shows the setup of the experiment. A small platinum resistance of FK222-1000-A with the size of 2.3 × 2.1 × 0.8 mm^3 was used as the temperature sensor and it was tightly glued to one side of the AMR sensor. Due to its small size, it had small influence to the temperature of the AMR sensor. The resistance of the Pt resistor was 1000 Ω at 0°C, the temperature coefficient was about 3850 ppm/K, and the temperature resolution was about 0.15°C. A DC current of 0.1 mA was used to bias the Pt resistor, and the voltage across the Pt resistor was amplified by an amplifier with the gain of 40 dB. The sample was fixed on the X-Y stage for the scanning. A computer was used to control the movement of the X-Y stage and to make the data acquisition and the data procession.

Figure 8 shows the temperature of the AMR sensor with different bias voltages from 0 to 33.4 V (a correction to [19], the bias voltage in [19] should be 2 times). For higher bias voltage, the thermal power of the AMR sensor was also bigger, so the temperature of the AMR sensor increased with the bias voltage. The temperature was 22.5°C in our laboratory. When

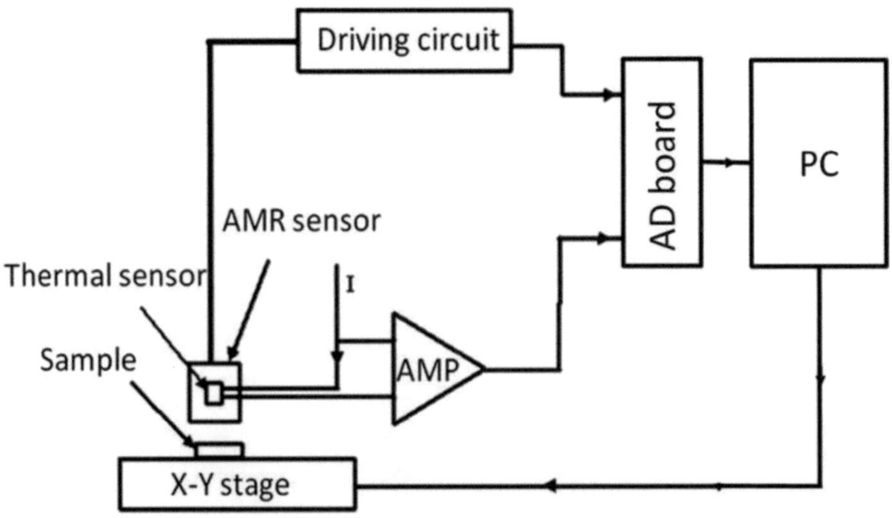

Figure 7. The experimental setup of simultaneously monitoring the temperature and the signal output of the AMR sensor.

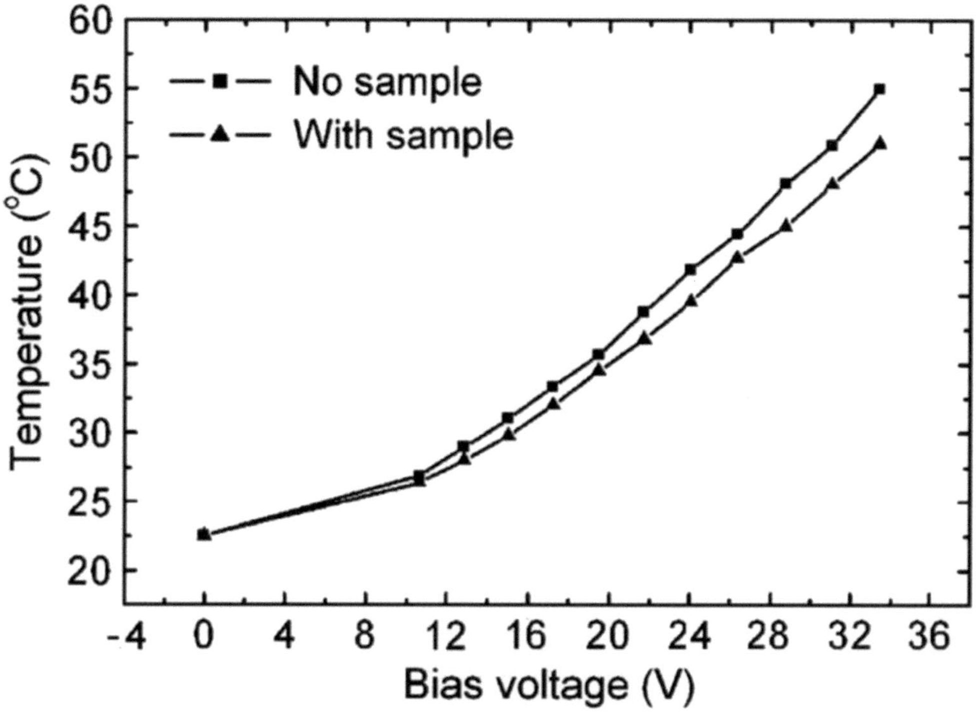

Figure 8. Temperature of the AMR sensor versus the bias voltage.

there was no sample, the temperature of the AMR sensor was about 42°C for the bias voltage of 24 V; when a 15 × 15 × 1 mm^3 copper plate was put under the AMR sensor with the lift-off distance of 0.2 mm, the temperature of the AMR sensor decreased to about 39.5°C.

For the AMR sensor of HMC1001, the temperature fluctuation causes the changes of the sensitivity, the resistance and the bridge offset, so the output signal of the AMR sensor also changes with the temperature of the AMR sensor. To measure the temperature fluctuation and the signal output of the AMR sensor correctly, we put a small copper plate with the size of about 15 × 15 × 1 mm^3 on the X-Y stage for the scanning. The lift-off distance between the AMR sensor and the copper plate was about 0.2 mm. The bias voltage of the AMR sensor was 24 V. The curves of (a) and (b) in **Figure 9** show the output signal and the temperature change of the AMR sensor when the copper plate was moved close to the AMR sensor. The two curves were very consistent, which proved that the output signal change of the AMR sensor was caused by the temperature fluctuation.

To reduce the influence of temperature drift and the heat transfer effect of AMR sensor, we developed a driving circuit for the AMR sensor with the Set/Reset method. **Figure 10** shows the block diagram. **Figure 11** shows the waveforms of the testing points in **Figure 10**. In our AMR sensor system, a square wave generator was used to produce a square wave with the frequency of about 20 kHz, which was the waveform 1 shown in **Figure 11**. After a power amplifier, a capacitor of 0.2 μF was used to produce the pulse to Set/Reset the AMR sensor.

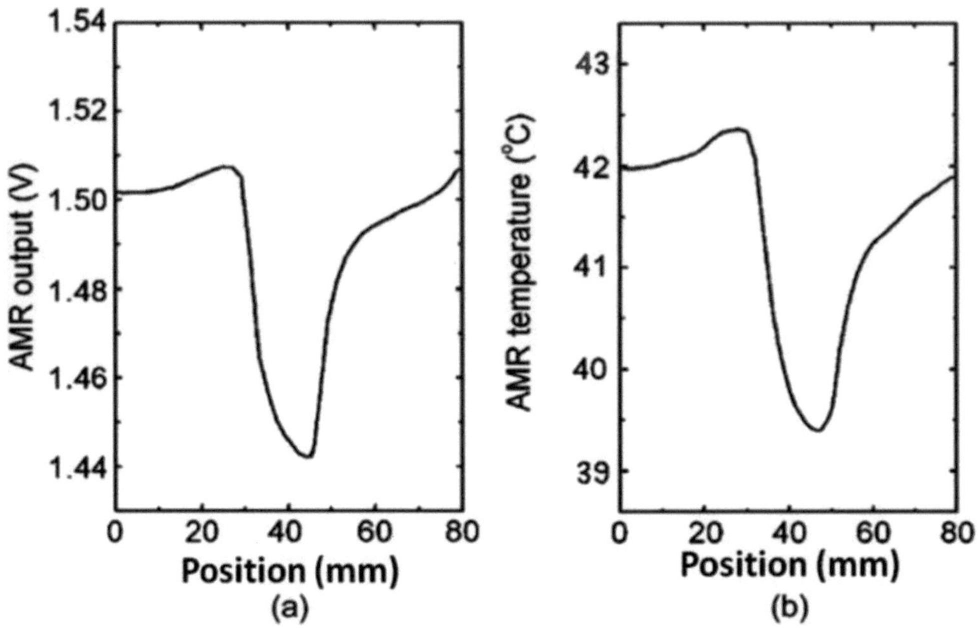

Figure 9. (a) Signal output of AMR sensor when a copper plate was close to the sensor. (b) Temperature change of the AMR sensor when the copper plate was close to the sensor.

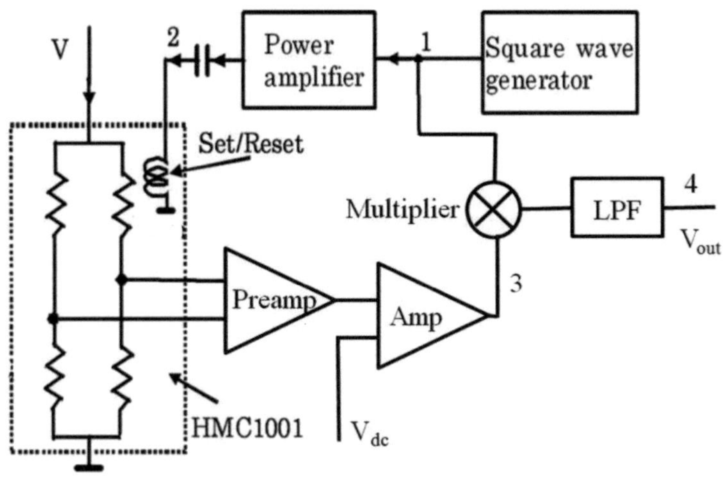

Figure 10. Block diagram of the AMR driving circuit with set/reset.

The pulse width was about 2 μs, which was the waveform 2 in **Figure 11**. The waveform 3 in **Figure 11** shows the output signal of the AMR bridge after a preamplifier with the gain of 40 dB. A multiplier was used to do the demodulation of the signal. After demodulation, a low-pass filter (LPF) with the cut-off frequency of 2 kHz was used and the waveform after the LPF was shown by waveform 4 in **Figure 11**.

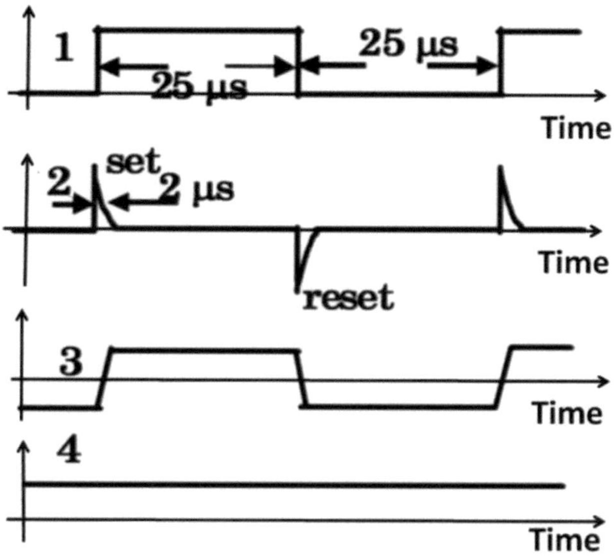

Figure 11. Waveforms of test point 1, 2, 3, and 4 in **Figure 10**.

We measured the output voltage signals for the AMR sensor with and without the Set/Reset method. The measurements were done in a magnetic shielded box. The measurement time was about 5 min with the measuring bandwidth from DC to 5 Hz. **Figure 12** shows the output

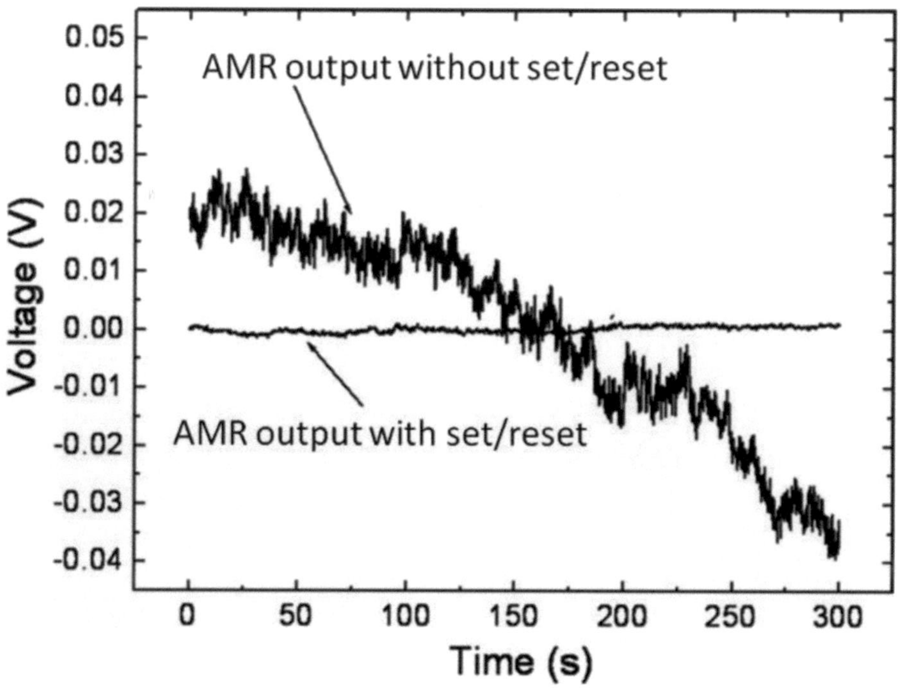

Figure 12. Output signal for the AMR sensor with Set/Reset and without Set/Reset.

signals. The total gains of the driving circuits for two methods were same. We observed that the output of the AMR sensor with the Set/Reset method was almost flat and the noise was smaller; however, the output without the Set/Reset method was not flat and the noise was bigger. The Set/Reset method was effective to reduce the influence of the variation of the temperature. Considering the total voltage/field transfer coefficient was about 6 mV/nT, the magnetic field resolution of the AMR sensor with the Set/Reset method was about 0.5 nT for the bandwidth from DC to 5 Hz.

3. Application to NDE

For the liquid rocket using liquid oxygen and liquid hydrogen, such as the H1-A rocket of Japan, the combustion chamber was made of Cu-Cr-Zr copper alloy (**Figure 13**). It was the ultra high temperature gas of about 3000 K inside of the combustion chamber. For cooling it, rectangular coolant passages were made in the wall of the cooper alloy and liquid hydrogen flew in them. With the very thin wall of 1 mm between the high temperature gas and the cooling liquid hydrogen, a big thermal gradient and an excessive thermal strain were generated in the inner wall of the combustion chamber. Due to the repeat of the oxidation/reduction, or the melting of the surface, some small cracks were generated. Now there was no effective method to detect the tiny defect in such complex structure. We were planning to detect the small defect in the copper alloy wall of combustion chamber using eddy current testing (ECT) with highly sensitive magnetic sensors.

We developed an ECT system using the AMR sensor and we successfully detected the artificial defects in plate-type samples [21, 22] and chamber-type samples [23], in which grooves and artificial defects were made to simulate the wall of the combustion chamber of a liquid fuel rocket. All experiments were done in an environment without shielding. **Figure 14** shows

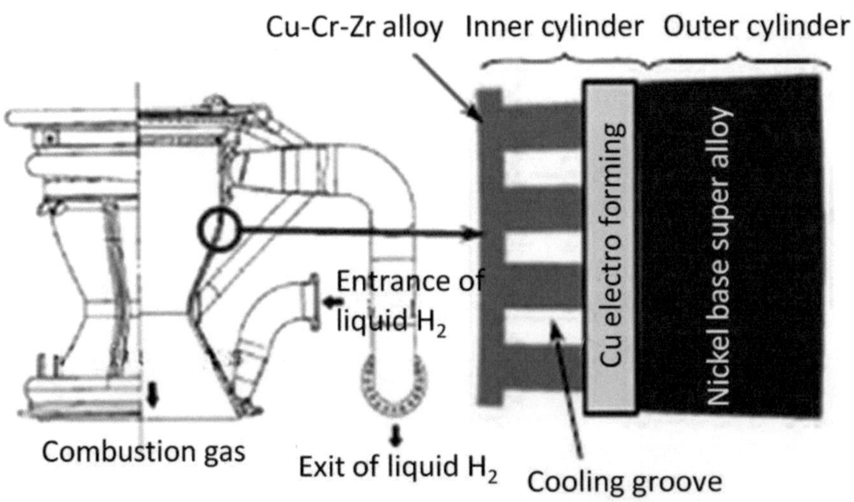

Figure 13. Combustion chamber of liquid rocket.

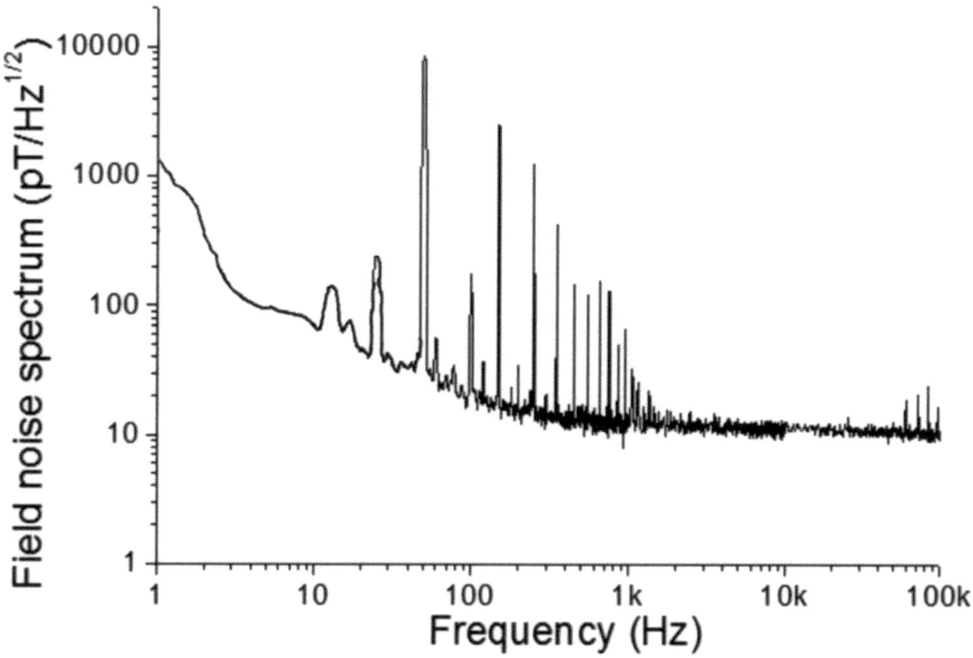

Figure 14. Magnetic field noise spectrum in our laboratory measured by AMR sensor.

the magnetic field noise spectrum measured in our laboratory using the AMR sensor. The peaks were 50 Hz interference and its harmonics.

First, we used an aluminium plate to simulate the complex structure of the inner wall of the liquid rocket combustion chamber. **Figure 15** shows the specimen. It was an aluminium plate with the size of 100 × 110 mm^2 and the thickness was 4 mm. Many parallel periodic ditches were made on one side of the aluminium plate. The depths of the ditches were 3 mm. The width of the ditches was 1 mm. Nine notches with different depths and lengths were made under the bottom of the ditches to simulate the defects of the combustion chamber.

Figure 16 shows the setup of the ECT system using the AMR sensor. The coil was used to produce the excitation field, and eddy current was induced in the specimen. The magnetic field produced by the eddy current was measured by the AMR sensor. The output signal of the AMR sensor was sent to a lock-in amplifier to get the amplitude signal or phase signal. The lock-in amplifier was also used to produce the sine signal sent to the excitation coil. In this AMR ECT system, the specimen was fixed on an X-Y stage. The movement of the X-Y stage was controlled by a computer. The movement resolution of the X-Y stage was about 50 μm, but the step of our measurement was set to 1 mm.

The excitation frequency was chosen according to the wall thickness between the scanning surface and the bottom of the notch. For the aluminium specimen, it was 1 mm, so 8 kHz was used, which had the penetration depth of about 1 mm. The current amplitude flow in the excitation coil was about 1 mA. The aluminium plate was fixed on the X-Y stage. The scanning was done over the backside of aluminium plate, where the ditches could not be seen. The scanning direction was perpendicular to the direction of crack.

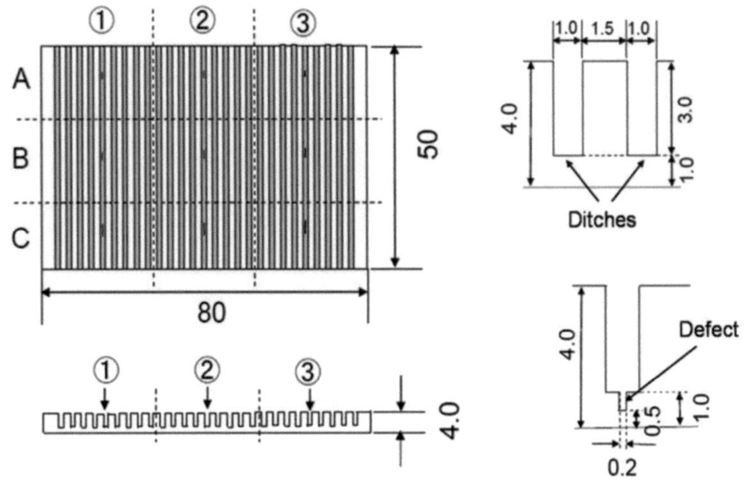

Figure 15. The aluminium specimen to simulate the combustion chamber and the defects. The depth and length of the notch defect under the bottom of some grooves: ①: depth of 0.25 mm. ②: depth of 0.5 mm. ③: depth of 0.75 mm. A: length of 2 mm. B: length of 5 mm. C: length of 10 mm.

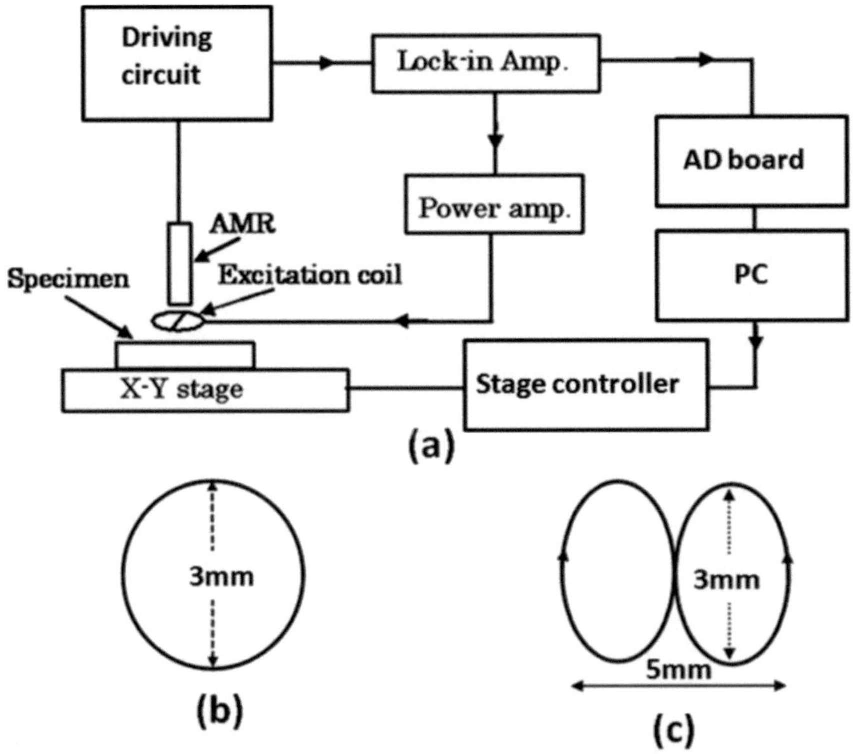

Figure 16. (a) The block diagram of the experimental setup of ECT using AMR sensor. (b) The circular excitation coil. (c) The double-D excitation coil.

Circular coil or differential coil, shown in **Figure 16(b)** and **(c)**, could be used as the excitation coil. When circular coil was used, the circular coil was 3 mm with 20 turns wound by 0.1 mm copper wire. The coil was attached to the bottom of the AMR sensor and the position of the coil was adjusted to make the output of the AMR biggest. When using circular coil, the strong background field was detected by the AMR sensor. When differential coil was used to produce the excitation field, the differential coil was 3 mm with 20 turns, which was attached to the bottom of the AMR sensor to produce the excitation field. When using differential coil, the background excitation field could be cancelled well. The position of the coil was adjusted to make sure that the AMR sensor had a smallest response to the excitation field. The other experimental conditions were same as that when using circular excitation coil.

Figure 17(a) shows the notch signals for one scanning using the circular coil, and **Figure 17(b)** shows the notch signals for one scanning using the differential coil. The signal-to-noise ratio was good for the cracks. **Figure 18** shows the 3D graph of the scanning results using the circular and **Figure 19** shows the 3D graph of the scanning results using the differential coil.

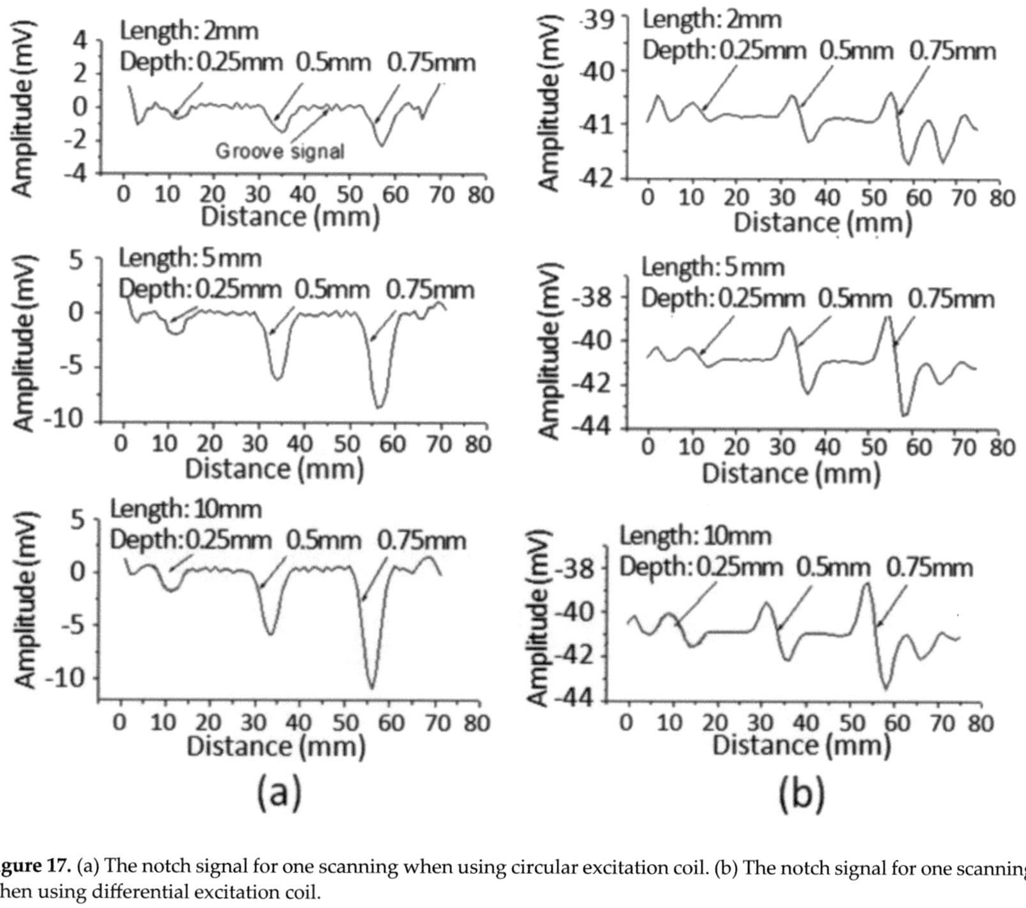

Figure 17. (a) The notch signal for one scanning when using circular excitation coil. (b) The notch signal for one scanning when using differential excitation coil.

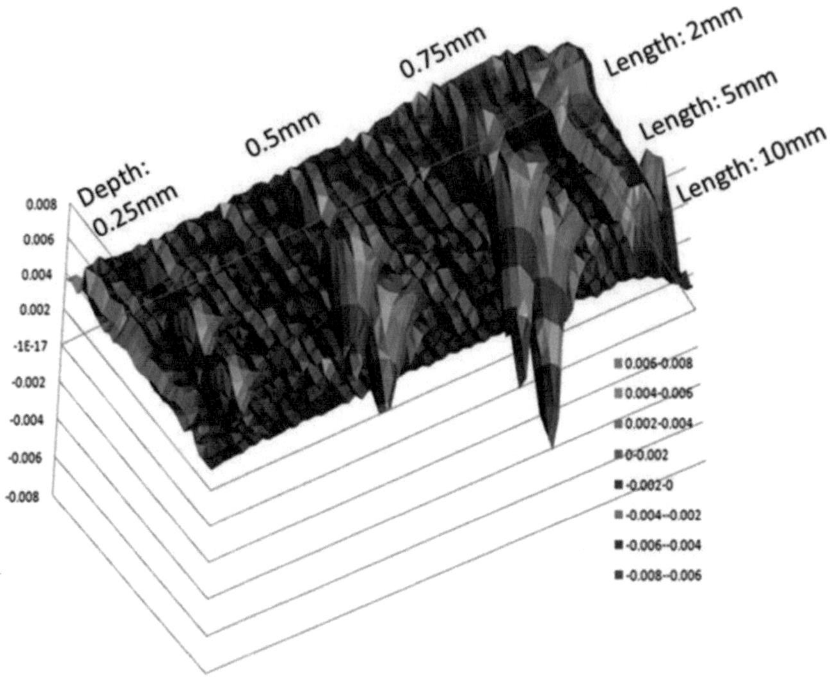

Figure 18. 3D graph of the results when using circular excitation coil.

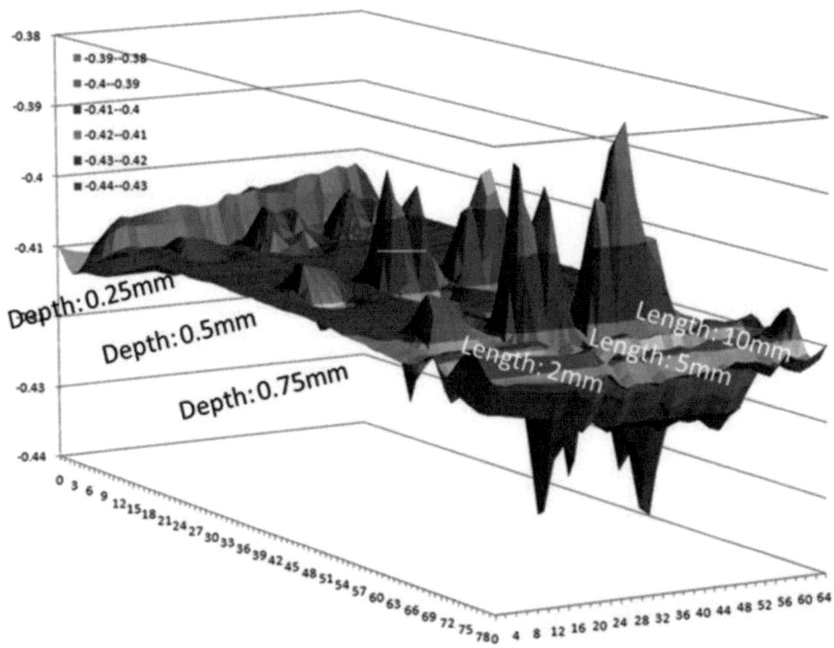

Figure 19. 3D graph of the results when using differential excitation coil.

We also prepared a chamber-type specimen made of copper alloy [23]. Grooves were fabricated to simulate the cooling grooves of the combustion chamber; and artificial notches were made under the bottoms of some grooves to simulate crack defects in the combustion chamber (**Figure 20**). The widths of the notches were about 0.3 mm. Notches with different lengths and depths were made. The lengths of the notches were 2, 5 and 10 mm, respectively; the remaining thickness of the wall at the positions of the notches was 0 mm (through), 0.2, 0.4 and 0.6 mm, respectively.

For chamber-type specimen, the lift-off variance during scanning might have a big influence to the output signals. Dual-frequency ECT [11, 23] was an effective method to reduce the influence of lift-off variance. **Figure 21** shows the schematic block diagram of the dual frequency ECT system with AMR sensor. The excitation coil was used to produce the AC magnetic field. Eddy currents were induced in the conductive specimen, and the AMR sensor was used to detect the magnetic field produced by the eddy currents. The sensing direction of the AMR sensor was along Z direction, which was perpendicular to the surface of the specimen. Two lock-in amplifiers and two frequencies were used in this ECT system. The amplitude output signal V_{f1} and V_{f2} of the two lock-in amplifiers were used and sent to a computer through an analog-to-digital (AD) board. Digital subtraction of V_{f1} and V_{f2} was done by the computer. The phase signals of the lock-in amplifiers were not used in our experiments.

The penetration depth of the applied AC magnetic fields could be estimated by the formula $\delta = (\pi f \mu \sigma)^{-1/2}$, where δ was the penetration depth, f was the frequency, μ was the magnetic permeability of the material and σ was the electrical conductivity of the material. For high frequency *f2*, the penetration depth was small, so mainly the surface roughness and variance of lift off were detected. For low frequency *f1*, the penetration depth was big; both the inside and surface properties of the material could be detected. If we subtracted the amplitude outputs (V1 and V2) of two lock-in amplifiers and choose a proper subtraction factor, it was possible to cancel the signals produced by the variance of liftoff during scanning.

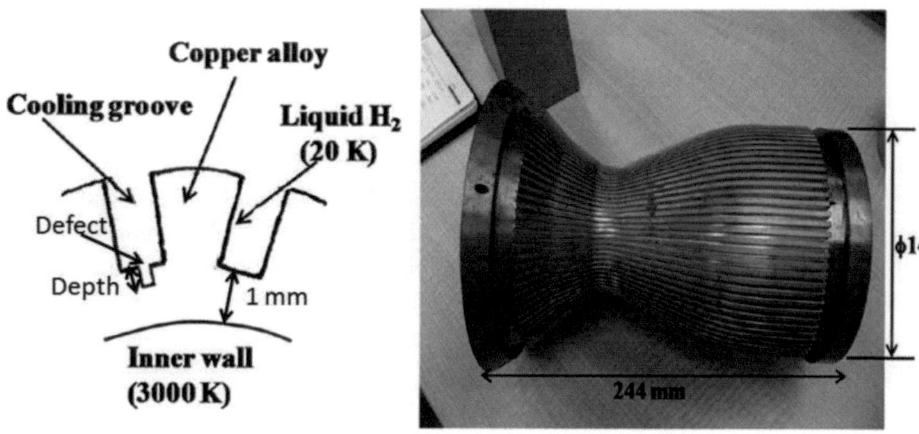

Figure 20. Chamber-type specimen to simulate the combustion chamber of liquid rocket.

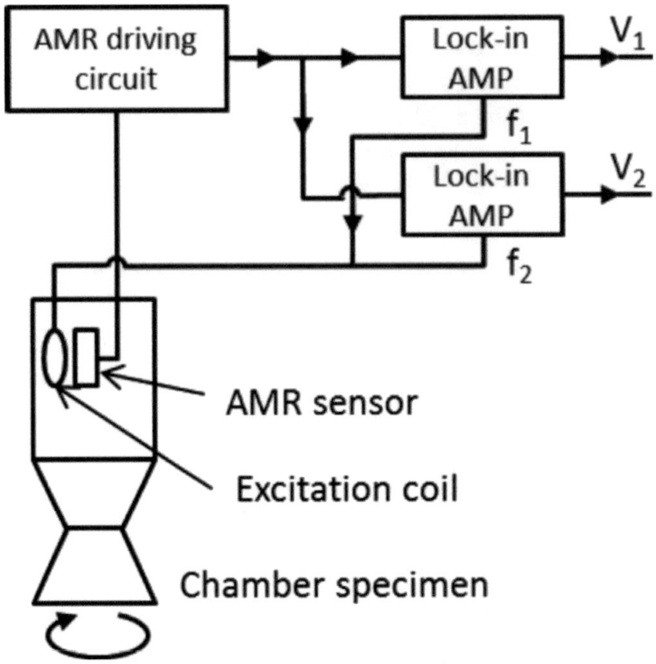

Figure 21. Dual frequency ECT system with AMR sensor.

The excitation coil was a 20 turn circular coil with an outer diameter of about 3 mm. The AMR sensor was located at the center of the excitation coil. The low frequency was 2 kHz and the high frequency was 20 kHz. The conductivity of the copper alloy was about 5×10^7 S/m. According to the wall thickness of the chamber (1 mm), 2 kHz was used as the low excitation frequency. The corresponding penetration depth was about 1.5 mm. The high excitation frequency was 20 kHz and the corresponding penetration depth was about 0.5 mm. The amplitude of the AC current flow in the excitation coil was about 10 mA. Scanning was realized by rotating the chamber with a motor. The rotating speed of the chamber was about 4 degree/s. The lift off between the surface of the specimen and the excitation coil was about 0.5 mm, and the variance of the lift off was about 0.2 mm during scanning.

Amplitude signals of the two lock-in amplifiers were used. **Figure 22** shows the scanning results for the notches with the remaining thicknesses of 0 mm (through) and 0.2 mm. Notches with the remaining thickness of 0 mm were detected by the low excitation frequency (2 kHz) and the high excitation frequency (20 kHz). For the notches with the remaining thickness of 0.2 mm, the signals became smaller for the high excitation frequency. During the scanning, the variance of the liftoff was about 0.2 mm, which causes the non-flatness of the lock-in amplifier output signals of $V_{2\,kHz}$ and $V_{20\,kHz}$. $V_{2\,kHz}-kV_{20\,kHz}$ was the subtraction result and k was a constant. For our experiments, when k = 0.46 was used, liftoff-related noise was compensated well. This proved that the dual frequency method was effective to reduce the influence of the variance of lift off.

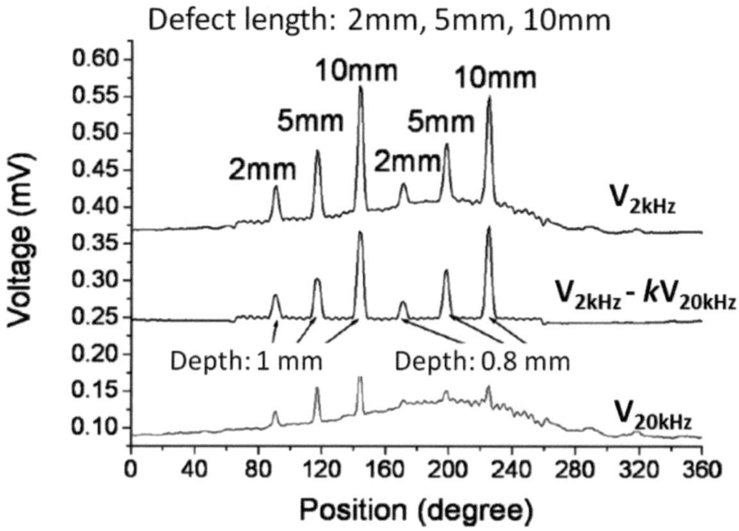

Figure 22. Scanning results for notches with the remaining wall thicknesses of 0 mm (through) and 0.2 mm.

Figure 23 shows the scanning results for the notches with the remaining wall thicknesses of 0.4 and 0.6 mm. For the low excitation frequency of 2 kHz, the defect signals could be obviously observed. For the high excitation frequency of 20 kHz, due to its small penetration depth, the notch signals were very small. Both the output signals of $V_{2\,kHz}$ and $V_{20\,kHz}$ were not flat, which was caused by the variance of the lift off during scanning. The subtraction result of $V_{2\,kHz}-kV_{20\,kHz}$ became flat, here k = 0.46 was also used. This also proved that the dual

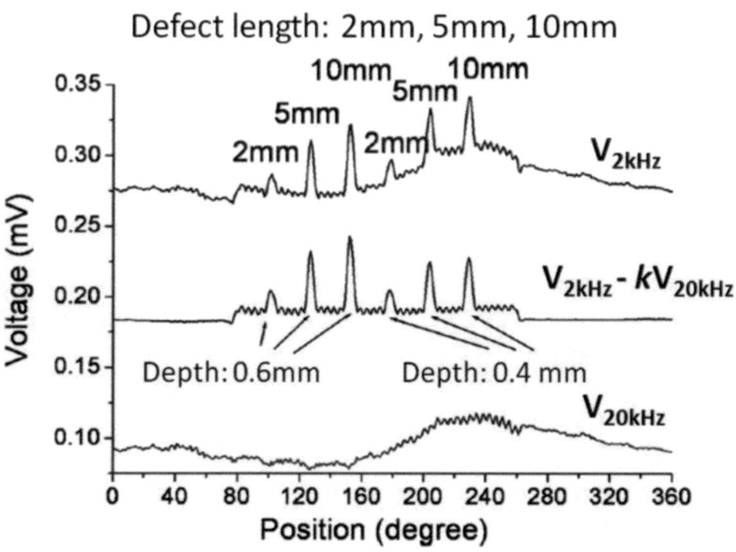

Figure 23. Scanning results for the notches with the remaining wall thicknesses of 0.4 and 0.6 mm.

frequency method was effective to reduce the influence of the variance of lift off. The small periodic signals were the signals produced by the grooves.

4. Conclusion

We improved the driving circuit of the AMR sensor. Using the commercial AMR sensor of HMC10001, lower magnetic field noise spectral densities of 12 pT/root(Hz) at 1 kHz and 20 pT/root(Hz) at 100 Hz were achieved. When the driving circuit operated at feedback mode, the distortion of the output was reduced. Using the Set/Reset method, the low frequency noise and the DC level drift caused by temperature effect were reduced. Using the optimized system of AMR sensor, we developed a high sensitive ECT NDE system and the small defects in the combustion chamber of liquid rocket were successfully detected.

Acknowledgements

We thank the helps from Dr M. Shiwa and Dr S. Moriya.

Author details

Dongfeng He

Address all correspondence to: he.dongfeng@nims.go.jp

National Institute for Materials Science, Tsukuba, Japan

References

[1] Tumanski S. Induction coil sensors—A review. Measurement Science and Technology. 2007;**18**:R31-R46

[2] Park DG, Angani CS, Rao BPC, Vertesy G, Lee DH, Kim KH. Detection of the subsurface cracks in a stainless steel plate using pulsed eddy current. Journal of Nondestructive Evaluation. 2013;**32**:350-353

[3] Bernieri A, Betta G, Ferrigno L, Laracca M. Improving performance of GMR sensors. IEEE Sensors Journal. 2013;**13**:4513-4521

[4] Betta G, Ferrigno L, Laracca M. GMR-based ECT instrument for detection and characterization of crack on a planar specimen: a hand-held solution. IEEE Transactions on Instrumentation and Measurement. 2012;**61**:505-512

[5] Jander A, Smith C, Schneider R. Magnetoresistive sensors for nondestructive evaluation. In: Conference on Advanced Sensor Technologies for Nondestructive Evaluation and Structural Health Monitoring; 8-10 March 2005; San Diego, CA. Proceedings of the Society of Photo-Optical Instrumentation Engineers (SPIE). Vol. 5770. 2005. pp. 1-13

[6] Allweins K, von Kreutzbruck M, Gierelt G. Defect detection in aluminium laser welds using an anisotropic magnetoresistive sensor array. Journal of Applied Physics. 2005;**97**:10Q102. DOI: 10.1063/1.1852391

[7] Koch RH, Rozen JR. Low-noise flux-gate magnetic-field sensors using ring and rod-core geometries. Applied Physics Letters. 2001;**78**:1897-1899

[8] Fujita Y, Sasada I. Detection of flaws in ferromagnetic samples based on low frequency eddy current imaging. Journal of Applied Physics. 2003;**93**:8277-8279

[9] He DF, Daibo M, Yoshizawa M. Mobile HTS rf SQUID magnetometer. IEEE Transactions on Applied Superconductivity. 2003;**13**:200-202

[10] Tavrin Y, krause HJ, Wolf W, Glyantsev V, Schubert J, Zander W, Bousack H. Eddy current technique with high temperature SQUID for non-destructive evaluation of non-magnetic metallic structures. Cryogenics. 1996;**36**:83-86

[11] He DF, Yoshizawa M. Dual-frequency eddy-current NDE based on high-T-c rf SQUID. Physica C: Superconductivity and Its Applications 2002;**383**:223-226

[12] Stutzke NA, Russek SE, Pappas DP. Low-frequency noise measurements on commercial magnetoresistive magnetic field sensors. Journal of Applied Physics. 2005;**97**:10Q107. DOI: 10.1063/1.1861375

[13] He DF, Tachiki M, Itozaki H. Highly sensitive anisotropic magnetoresistance magnetometer for eddy-current nondestructive evaluation. Review of Scientific Instruments. 2009;**80**:036102. DOI: 10.1063/1.3098946

[14] Thomson W. On the electrodynamic qualities of metals: Effects of magnetization on the electric conductivity of nickel and iron. Proceedings of the Royal Society of London. 1857;**8**:546-550

[15] McGuire TR, Potter RI. Anisotropic magnetoresistance in ferromagnetic 3D alloys. IEEE Transactions on Magnetics. 1975;**11**:1018-1037

[16] Kuijk KE, Vangestel WJ, Gorter FW. Barber pole, a linear magnetoresistive head. IEEE Transactions on Magnetics. 1975;**11**:1215-1217. DOI: 10.1109/TMAG.1975.1058886

[17] The Datasheet of HMC1001. Honeywell International Inc, https://aerospace.honeywell.com/en/~/media/aerospace/files/datasheet/1and2axismagneticsensorshmc1001-1002-1021-1022_ds.pdf

[18] West FG. Magnetoresistive measurements on domain rotation nickel-iron alloy films. Nature. 1960;**188**:129-130

[19] He DF, Jia JP. Heat transfer effects in scanning anisotropic magnetoresistive imaging. IEEE Transactions on Magnetics. 2011;**47**:264-266. DOI: 10.1109/TMAG.2010.2088131

[20] He DF, Shiwa M. An anisotropic magneto resistive sensor with set/reset field. Review of Scientific Instruments. 2011;**82**:094703. DOI: 10.1063/1.3640412

[21] He DF, Shiwa M, Uetake I, Yamawaki H, Tachiki M, Itozaki H. Basic examination for the defect detection of combustion chamber using ECT with AMR sensor. Journal of JSNDI. 2010;**59**:510-514

[22] He DF, Shiwa M, Jia JP, Takatsubo J, Moriya S. Multi-frequency ECT with AMR sensor. NDT & E International. 2011;**44**:438-441. DOI: 10.1016/j.ndteint.2011.04.004

[23] He DF, Zhang YZ, Shiwa M, Moriya S. Development of eddy current testing system for inspection of combustion chambers of liquid rocket engines. Review of Scientific Instruments. 2013;**84**:014701. DOI: 10.1063/1.4773539